Jianguo Zhang, Ling Shao, Lei Zhang, and Graeme A. Jones (Eds.)
Intelligent Video Event Analysis and Understanding

Studies in Computational Intelligence, Volume 332

Editor-in-Chief
Prof. Janusz Kacprzyk
Systems Research Institute
Polish Academy of Sciences
ul. Newelska 6
01-447 Warsaw
Poland
E-mail: kacprzyk@ibspan.waw.pl

Further volumes of this series can be found on our homepage: springer.com

Vol. 312. Patricia Melin, Janusz Kacprzyk, and Witold Pedrycz (Eds.)
Soft Computing for Recognition based on Biometrics, 2010
ISBN 978-3-642-15110-1

Vol. 313. Imre J. Rudas, János Fodor, and Janusz Kacprzyk (Eds.)
Computational Intelligence in Engineering, 2010
ISBN 978-3-642-15219-1

Vol. 314. Lorenzo Magnani, Walter Carnielli, and Claudio Pizzi (Eds.)
Model-Based Reasoning in Science and Technology, 2010
ISBN 978-3-642-15222-1

Vol. 315. Mohammad Essaaidi, Michele Malgeri, and Costin Badica (Eds.)
Intelligent Distributed Computing IV, 2010
ISBN 978-3-642-15210-8

Vol. 316. Philipp Wolfrum
Information Routing, Correspondence Finding, and Object Recognition in the Brain, 2010
ISBN 978-3-642-15253-5

Vol. 317. Roger Lee (Ed.)
Computer and Information Science 2010
ISBN 978-3-642-15404-1

Vol. 318. Oscar Castillo, Janusz Kacprzyk, and Witold Pedrycz (Eds.)
Soft Computing for Intelligent Control and Mobile Robotics, 2010
ISBN 978-3-642-15533-8

Vol. 319. Takayuki Ito, Minjie Zhang, Valentin Robu, Shaheen Fatima, Tokuro Matsuo, and Hirofumi Yamaki (Eds.)
Innovations in Agent-Based Complex Automated Negotiations, 2010
ISBN 978-3-642-15611-3

Vol. 320. xxx

Vol. 321. Dimitri Plemenos and Georgios Miaoulis (Eds.)
Intelligent Computer Graphics 2010
ISBN 978-3-642-15689-2

Vol. 322. Bruno Baruque and Emilio Corchado (Eds.)
Fusion Methods for Unsupervised Learning Ensembles, 2010
ISBN 978-3-642-16204-6

Vol. 323. Yingxu Wang, Du Zhang, and Witold Kinsner (Eds.)
Advances in Cognitive Informatics, 2010
ISBN 978-3-642-16082-0

Vol. 324. Alessandro Soro, Vargiu Eloisa, Giuliano Armano, and Gavino Paddeu (Eds.)
Information Retrieval and Mining in Distributed Environments, 2010
ISBN 978-3-642-16088-2

Vol. 325. Quan Bai and Naoki Fukuta (Eds.)
Advances in Practical Multi-Agent Systems, 2010
ISBN 978-3-642-16097-4

Vol. 326. Sheryl Brahnam and Lakhmi C. Jain (Eds.)
Advanced Computational Intelligence Paradigms in Healthcare 5, 2010
ISBN 978-3-642-16094-3

Vol. 327. Slawomir Wiak and Ewa Napieralska-Juszczak (Eds.)
Computational Methods for the Innovative Design of Electrical Devices, 2010
ISBN 978-3-642-16224-4

Vol. 328. Raoul Huys and Viktor K. Jirsa (Eds.)
Nonlinear Dynamics in Human Behavior, 2010
ISBN 978-3-642-16261-9

Vol. 329. Santi Caballé, Fatos Xhafa, and Ajith Abraham (Eds.)
Intelligent Networking, Collaborative Systems and Applications, 2010
ISBN 978-3-642-16792-8

Vol. 330. Steffen Rendle
Context-Aware Ranking with Factorization Models, 2010
ISBN 978-3-642-16897-0

Vol. 331. Athena Vakali and Lakhmi C. Jain (Eds.)
New Directions in Web Data Management 1, 2011
ISBN 978-3-642-17550-3

Vol. 332. Jianguo Zhang, Ling Shao, Lei Zhang, and Graeme A. Jones (Eds.)
Intelligent Video Event Analysis and Understanding, 2011
ISBN 978-3-642-17553-4

Jianguo Zhang, Ling Shao, Lei Zhang,
and Graeme A. Jones (Eds.)

Intelligent Video Event Analysis and Understanding

Springer

Dr. Jianguo Zhang
School of Computing
University of Dundee
Dundee DD1 4HN
Scotland, UK
E-mail: jgzhang@computing.dundee.ac.uk
http://www.computing.dundee.ac.uk/staff/jgzhang/

Dr. Ling Shao
Department of Electronic & Electrical Engineering
The University of Sheffield, Sheffield, S1 3JD, UK
E-mail: ling.shao@sheffield.ac.uk
http://lshao.staff.shef.ac.uk/

Dr. Lei Zhang
Lead Researcher, Microsoft Research Asia,
49 Zhichun Road, Beijing 100190, P.R. China
E-mail: leizhang@microsoft.com
http://research.microsoft.com/en-us/people/leizhang/

Prof. Graeme A. Jones
Digital Imaging Research Centre,
Faculty of Computing, Information Systems and Mathematics,
Kingston University, Penrhyn Road, Kingston upon Thames, Surrey KT1 2EE, UK
E-mail: g.jones@kingston.ac.uk
http://cism.kingston.ac.uk/people/details.php?AuthorID=3

ISBN 978-3-642-17553-4   e-ISBN 978-3-642-17554-1

DOI 10.1007/978-3-642-17554-1

Studies in Computational Intelligence   ISSN 1860-949X

© 2011 Springer-Verlag Berlin Heidelberg

This work is subject to copyright. All rights are reserved, whether the whole or part of the material is concerned, specifically the rights of translation, reprinting, reuse of illustrations, recitation, broadcasting, reproduction on microfilm or in any other way, and storage in data banks. Duplication of this publication or parts thereof is permitted only under the provisions of the German Copyright Law of September 9, 1965, in its current version, and permission for use must always be obtained from Springer. Violations are liable to prosecution under the German Copyright Law.

The use of general descriptive names, registered names, trademarks, etc. in this publication does not imply, even in the absence of a specific statement, that such names are exempt from the relevant protective laws and regulations and therefore free for general use.

Typeset & Cover Design: Scientific Publishing Services Pvt. Ltd., Chennai, India.

Printed on acid-free paper

9 8 7 6 5 4 3 2 1

springer.com

Preface

With the vast development of Internet capacity and speed, as well as wide adoption of media technologies in people's daily life, a large amount of videos have been surging, and need to be efficiently processed or organized based on interest. The human visual perception system could, without difficulty, interpret and recognize thousands of events in videos, despite high level of video object clutters, different types of scene context, variability of motion scales, appearance changes, occlusions and object interactions. For a computer vision system, it has been be very challenging to achieve automatic video event understanding for decades. Broadly speaking, those challenges include robust detection of events under motion clutters, event interpretation under complex scenes, multi-level semantic event inference, putting events in context and multiple cameras, event inference from object interactions, etc.

In recent years, steady progress has been made towards better models for video event categorisation and recognition, e.g., from modelling events with bag of spatial temporal features to discovering event context, from detecting events using a single camera to inferring events through a distributed camera network, and from low-level event feature extraction and description to high-level semantic event classification and recognition. Nowadays, text based video retrieval is widely used by commercial search engines. However, it is still very difficult to retrieve or categorise a specific video segment based on their content in a real multimedia system or in surveillance applications. To advance the progress further, we must adapt recent or existing approaches to find new solutions for intelligent video understanding.

This book aims to present state-of-the-art research advances of video event understanding technologies. It will provide researchers and practitioners a rich resource for future research directions and successful practice. It could also serve as a reference tool and handbook for researchers in a number of applications including visual surveillance, human-computer interaction, and video search and indexing etc. Its potential audience will be composed of active researchers and practitioners as well as graduate students working on video analysis in various disciplines such as computer vision, pattern recognition, information security, artificial intelligence, etc.

In Chapter 1, Vatavu addresses a double view of understanding meaningful events in gesture based interaction: events that specify gestures together with intelligent algorithms that detect them in video sequences; gestures, that once recognized and accordingly interpreted by the system, become important events in the human-computer dialogue specifying the common understanding that was established. The chapter follows the duality aspect of events from the system as well as the human perspective contributing to the present understanding of gestures in human-computer interaction.

In Chapter 2, Yu and Zhang present a motion segmentation approach based on the subspace segmentation technique, the generalized PCA. By incorporating the cues from the neighbourhood of intensity edges of images, motion segmentation is solved under an algebra framework. They propose an effective post-processing procedure, which can detect the boundaries of motion layers and further determine the layer ordering.

In Chapter 3, Zhou presents a strategy based on human gait to achieve efficient tracking, recovery of ego-motion and 3-D reconstruction from an image sequence acquired by a single camera attached to a pedestrian. In the first phase, the parameters of the human gait are established by a classical frame-by-frame analysis, using a generalised least squares (GLS) technique. In the second phase, this gait model is employed within a ``predict-correct'' framework using a maximum a posterior, expectation maximization (MAP-EM) strategy to obtain robust estimates of the ego-motion and scene structure, while continuously refining the gait model.

In Chapter 4, Mattivi and Shao apply the Local Binary Pattern on Three Orthogonal Planes (LBP-TOP) descriptor to the field of human action recognition. They use LBP and CS-LBP techniques combined with gradient and Gabor images. Several modifications and extensions to the descriptor are further developed.

In Chapter 5, Zhuang et al. present an efficient object localization approach based on the Gaussianized vector representation following a branch-and-bound search scheme introduced by Lampert et al. In particular, they design a quality bound for rectangle sets characterized by the Gaussianized vector representation for fast hierarchical search. Further, they propose incorporating a normalization approach that suppresses the variation within the object class and the background class. This method outperforms previous work using the histogram-of-keywords representation for object localisation.

In Chapter 6, Zhang and Gong present a framework for robust people detection in highly cluttered scenes with low resolution image sequences. Their model utilises both human appearance and their long-term motion information. Preliminary studies demonstrate the method achieved good results under challenging conditions.

In Chapter 7, Hervieu and Bouthemy describe object-based approach for temporal analysis of sports videos using player's trajectories. An original hierarchical parallel semi-Markov model (HPaSMM) is proposed. Such probabilistic graphical models help taking into account low level temporal causalities of trajectories features as well as upper level temporal transitions between activity phases. It can be used for applications of sports video semantic-based understanding such that segmentation, summarization and indexing.

In Chapter 8, Davis et al. describe an experimental system for the recognition of human faces from surveillance video. Their system detects faces using the Viola-Jones face detector, and then extracts local features to build a shape-based feature vector. Consideration was given to improving the performance and accuracy of both the detection and recognition steps.

In Chapter 9, Odashima et al. propose an object movement detection method in a household environment via the stable changes of images. To detect object placements and object removals robustly, the method employs the layered background model and the edge subtraction based classification method. In addition, to classify objects and non-objects robustly though the changed regions are occluded, the method uses motion history of the regions.

In Chapter 10, Ali et al. provide a survey on BBC Dirac Video Codec which can be use for compressing high resolution files, broadcasting, live video streaming, pod casting, and desktop production. This survey not only provides an in-deep description of different version of Dirac Video Codec but also explain the algorithmic explanation of Dirac at implementation level. It aims to help to new researchers who are working to understand BBC Dirac video codec but also provide them future directions and ideas to enhance features of BBC Dirac video codec.

Editors

Jianguo Zhang
Ling Shao
Lei Zhang
Graeme A. Jones

Contents

The Understanding of Meaningful Events in Gesture-Based
Interaction 1
Radu-Daniel Vatavu

Apply GPCA to Motion Segmentation....................... 21
Hongchuan Yu, Jian J. Zhang

Gait Analysis and Human Motion Tracking 39
Huiyu Zhou

Spatio-temporal Dynamic Texture Descriptors for Human
Motion Recognition 69
Riccardo Mattivi, Ling Shao

Efficient Object Localization with Variation-Normalized
Gaussianized Vectors 93
*Xiaodan Zhuang, Xi Zhou, Mark A. Hasegawa-Johnson,
Thomas S. Huang*

Fusion of Motion and Appearance for Robust People
Detection in Cluttered Scenes 111
Jianguo Zhang, Shaogang Gong

Understanding Sports Video Using Players Trajectories 125
Alexandre Hervieu, Patrick Bouthemy

Real-Time Face Recognition from Surveillance Video 155
Michael Davis, Stefan Popov, Cristina Surlea

Event Understanding of Human-Object Interaction: Object Movement Detection via Stable Changes 195
Shigeyuki Odashima, Taketoshi Mori, Masamichi Simosaka, Hiroshi Noguchi, Tomomasa Sato

Survey of Dirac: A Wavelet Based Video Codec for Multiparty Video Conferencing and Broadcasting............ 211
Ahtsham Ali, Nadeem A. Khan, Shahid Masud, Syed Farooq Ali

Author Index 249

The Understanding of Meaningful Events in Gesture-Based Interaction

Radu-Daniel Vatavu

Abstract. Gesture-based interaction is becoming more and more available each day with the continuous advances and developments in acquisition technology and recognition algorithms as well as with the increasing availability of personal (mobile) devices, ambient media displays and interactive surfaces. Vision-based technology is the preferred choice when non-intrusiveness, unobtrusiveness and comfortable interactions are being sought. However, it also comes with the additional costs of difficult (unknown) scenarios to process and far-than-perfect recognition rates. The main challenge is represented by spotting and segmenting gestures in video media. Previous research has considered various events that specify when a gesture begins and when it ends in conjunction with location, time, motion, posture or various other segmentation cues. Therefore, video events identify, specify and segment gestures. Even more, when gestures are being correctly detected and recognized by the system with the appropriate feedback delivered to the human, the result is that gestures become themselves events in the human-computer dialogue: the commands were understood and the system reacted back.

This chapter addresses the double view of meaningful events: events that specify gestures together with intelligent algorithms that detect them in video sequences; gestures, that once recognized and accordingly interpreted by the system, become important events in the human-computer dialogue specifying the common understanding that was established. The chapter follows the duality aspect of events from the system as well as the human perspective contributing to the present understanding of gestures in human-computer interaction.

Radu-Daniel Vatavu
University Stefan cel Mare of Suceava, 13 Universitatii,
720229 Suceava, Romania
e-mail: vatavu@eed.usv.ro

1 Introduction

Gestures in interfaces are becoming more and more popular as the advances in technology have been allowing proliferation of software and devices with such interaction capabilities [40]. An increasing interest may be observed in adopting such interfaces with a great mass of consumers owning personal mobile devices that allow entering strokes via a stylus or simply using touch-based commands. The same interest is equally observed in various research communities concerned with gesture acquisition and analysis such as human-computer interaction, computer vision or pattern recognition [12, 33, 34, 38, 45]. Although the technology is clearly at the beginning (as there is a huge difference between entering strokes on a mobile device and interacting freely with computer systems or ambient media via natural gesturing), the motivation that drives all these developments is a strong one: gestures are familiar, easy to perform and natural to understand. The common perception is that an interface capable of correctly understanding and interpreting human gestures would be ideal by allowing similar interactions in the human-computer dialogue as they appear in our every-day life.

A great number of technologies are available today for capturing human input and video cameras represent one of them. The advantage that computer vision brings with respect to the other technologies is that users are not bound to any equipment: there is nothing to be worn or held that could importunate or burden the interaction. This encourages free hand gestures, natural head movements or even whole body input in accordance with the application or the task needs. The downside of this great flexibility in interaction is represented by the amount of computations required for processing video as well as by the current state of computer vision technology, a field of research that is still in its dawn with far-than-perfect recognition rates. One way to get around these inconveniences is to control the working scenario but sometimes this proves challenging if not impossible (as it would be the case of a mobile robot capable of interpreting gestures but which travels freely outside the laboratory). The other idea is to use specific video events that can help determining when a gesture started and when it ended by event-based detection, reasoning and inferring with simple yet robust assumptions.

This chapter proposes a discussion on events that help the process of detecting gestures in video sequences by identifying and describing the most frequent approaches known to work for such a task. Criteria such as location, time, posture or various combinations have been used in order to infer both on the users as well as on the gestures they perform. A gesture is thus specified by various predefined *events* that are being detected or monitored in the video sequence. Equally important, the chapter takes into consideration what happens from the user's point of view: once a gesture has been successfully detected, recognized, and its associated action executed accordingly, the feedback the users are receiving proves extremely important for the fluency of the interaction. From the users' perspective, a recognized gesture that has been acknowledged appropriately represents an important *event* in the new established communication as it would be in the general human-human dialogue. This can be explained by humans' predisposition and willingness to anthropomorphize unanimated objects they are interacting with hence by attributing human-computer interactions the same characteristics that a human-human dialogue or communication would posses. For example, according to Sharma and Kurian [46]:

> "Verbal and non-verbal human behavior form the events for human communication. Verbal behavior makes use of language, while the nonverbal behavior employs body postures, gestures and actions of different kinds."

The same verbal and non-verbal behavior are employed by humans whether they are having as interaction partner another human or a computer system. Therefore, it can be expected that the feedback received from the computer would also be interpreted as a communication event. The experience of previous human-human communications is used unconsciously for human-computer interaction for which the feedback mechanisms are interpreted in a similar manner. Figure 1 illustrates the different expressions and roles of events in the human-computer interaction cycle.

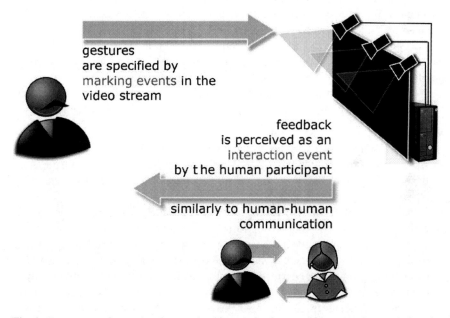

Fig. 1 Gestures are detected and recognized by monitoring specific events that mark them in the video stream while the appropriate feedback of a recognized gesture is perceived as an interaction event by the human participant.

The motivation behind the gesture-event relationship is thus introduced and discussed: events define and segment gestures while correctly identified gestures represent themselves events in the human-computer interaction cycle. Section 2 of the chapter describes techniques which are commonly used for spotting gestures in video frames by using event-based monitoring while section 3 brings evidence with regards to the feedback mechanisms that transform into events in the human-computer dialogue. The main purpose of the chapter is thus to bring further knowledge to the current understanding of gesture-based interfaces by the means of meaningful events and by looking at the topic from two different but equally important perspectives.

2 Events for Spotting and Detecting Gestures

Techniques for spotting gestures in video sequences are being implemented by detecting and monitoring (or tracking) custom events defined using location, time and posture criteria. The successful detection of one event triggers gesture recording while detection of another starts the gesture recognizer that classifies the recorded motion. This section analyses various event types and their associated criteria as well as the algorithms and techniques employed for segmenting gestures from sequences of continuous motion. The focus will be primarily oriented towards vision-based computing but other capture technologies will be mentioned when the techniques and algorithms used for segmenting human motions are relevant to the discussion. Jaimes and Sebe [45], Poppe [38], Moeslund et al. [33, 34] and Erol et al. [12] provide extensive surveys on vision-based motion capture and analysis as well as on multimodal human-computer interaction and they represent good starting points for a general overview on the advances in these fields.

We identify four different event types that have been used extensively either singly or in various combinations for segmenting gestures in video sequences:

- *Location* represents a powerful cue for detecting postures as well as for segmenting interesting gestures from continuous movements. Requiring that a gesture starts or ends in a predefined region or knowing/learning that some locations in the scene are more likely to contain valid gestures leads to great reduction in algorithmic complexity;
- *Posture* information allows marking gesture commands in a way that feels natural and accessible for the users to perform and model cognitively: for example, a given posture could mark the beginning of a gesture while another signals its ending. Posture is a robust cue that gives both the user as well as the system the certainty that a gesture command is being entered: the system is able to filter out the majority of movements while being interested only in the actual gesture commands. Also, users are creating themselves a mental model for the interaction process: commands are issued only if specified postures are being executed in a similar manner to how click-like events work in standard WIMP interfaces;
- *Tap and touch* events can be detected by touch-sensitive materials as well as by video cameras (most horizontal interactive surfaces use IR video cameras in order to detect touch events on the tabletop). A tap or a touch is clearly perceived as a marking event from both the system as well as the user's perspective. Touching clearly signifies both intent as well as command during the interaction process;
- *Custom*-based events other than the above may be additionally used in order to ease even further the gesture detection process. They usually relate to various

equipments that are being held or worn and which provide additional levels of accuracy and precision for the interaction. Hand tension or muscle activity represent such examples [16].

Figure 2 visually illustrates the four events in conjunction with their relationship to various information categories they support. For example, location events can identify *where* the action takes place, *when* it took place but also *who* performed it by position reasoning and inferring. Posture events specify *when* a gesture started, when it ended, whether it passed through some intermediate state as well as *what* that gesture represented. Even more, executing a given posture is related to *intention* to interact which is clearly understood by both the system and user. Tap and touch allow identification of *where* and *when* the action took place and they are also strongly connected to *intention* (touching clearly specifies interest, designation and willingness for interaction from the user's perspective). Custom events may be specifically designed in order to address any of these questions.

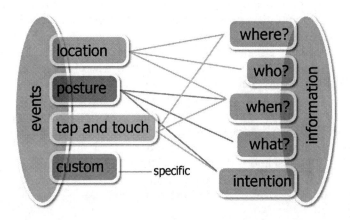

Fig. 2 Event types in video processing (location, posture, tap, touch and custom events) and their relationships to the information categories they address.

2.1 Location-Based Events

Location criteria can be used in order to define regions of interest in the video sequence to be processed where hands or other body parts are likely to be positioned at the start, during or at the end of a gesture. The advantages are multiple:

- First of all, the amount of image processing that is required is reduced only to the most promising regions that will likely contain body parts engaged in gesture production;
- The accuracy of the overall procedure is likely to become higher as the rest of the image is simply ignored which considerably reduces the chance of outputting

false positives. This is important in unknown scenarios where color or motion-based reasoning may fail cause of complex backgrounds to analyze (such as crowded scenes with multiple motion sources or objects that present the same color as the human skin);
- Systems that employ general tracking algorithms usually require an initialization stage during which the object to be tracked needs to be presented to the system. A special location or a combination between location and posture are usually employed in order to initialize such trackers;
- The segmentation process of a gesture is considerably facilitated if the start and ending locations are known in advance. Segmenting gestures from continuous movements is a difficult task hence knowing where and when a gesture started and when it ended by simple location event triggers is definitely a plus on system robustness.

Location-based events are triggered whenever motion, color or other features are identified in such predefined regions of interest. They may specify when the gesture started, when it ended or whether it passed through some intermediate stage towards completion. If the scenario is known and the location fixed and static (for example when working at a desk with the video camera monitoring the tabletop as Figure 3 illustrates), the regions may be simply defined during first-time installation and calibration of the system. For example, Vatavu and Pentiuc [50] use such location-based events in order to detect and track hand postures only when hands are positioned in predefined regions on the surface of the interactive coffee table prototype. As the scenario is a static one (or unlikely to change much since installation), the calibration procedure represented by identifying and defining the regions of interest is promising to produce robust results.

However, if the scenario is dynamic and likely to change over time (e.g. what a mobile robot will see during its travel), the common solution is to use a robust technique that would identify some aspect of the human body and then monitor

Fig. 3 Static scenario corresponding to a fixed camera installation for which regions of interest were defined during system setup and calibration. Left: processing can be reduced to limited regions only and even hand-wise delimitation is possible. Right: special regions such as the keyboard (1) or printer (4) areas could trigger special events.

regions of interest around it. The preferred choice in this case would be to detect the operator's face as robust solutions such as the Viola and Jones face detector do exist for accomplishing this task [52]. Once the location of the operator face and, implicitly, the location of his head are known, the positions of other body parts are inferred using simple geometry constraints. Figure 4 illustrates this idea with several regions dynamically computed around the user's head at one arm-length around his body. Even more, detecting the face brings in another advantage: as the face region will expose skin-colored pixels, on-the-fly calibration of color detectors can be performed [7] which increases system robustness and adaptability to the environment.

Fig. 4 The user's face can be robustly detected and used subsequently for inferring the locations of other body parts. Following [7] and [32], special regions of interest may be defined around the bounding rectangle of the user's face. When the hand is detected in one particular region an event will be triggered signaling the start, continuation or ending of a gesture. A wave gesture can be described for example as a series of 1-2-1-2-1 location events.

Cerlinca et al. [7] define and use such regions of interest around the human body in order to facilitate segmentation and recognition of free hand gestures. Their purpose is to transmit commands to a mobile robot that may circulate in any environment. Ten such regions are defined around the operator's face which is reliably detected using the Viola and Jones classifier. Skin color-based thresholding is further applied to these regions only in order to detect the operator's hands. Combinations of several active regions (for which hands were detected inside) correspond to various gesture commands such as move forward, backward, turn left, turn right, etc.

Marcel [32] performs a similar partitioning of the space around the user's face combined with anthropometry constraints in order to detect the location of the hand. The scenario is very similar to the one illustrated in Figure 4. The user's intention for producing a gesture is detected by monitoring the active windows that are formed in the body-face space. When skin-colored objects are detected within these predefined windows, assumption of the user hand is taken and a posture recognizer further employed.

Iannizzotto et al. [19] use the same principle of location-based events for detecting when a gesture command starts. The authors' Graylevel VisualGlove system does not track the whole hand but rather the thumb and index fingers only in order to simulate mouse operations. When both fingers are detected inside a predefined area (the initialization stage), the tracking procedure begins and actions are being recognized. Also, whenever tracking is lost, the initialization procedure must be repeated with the fingers positioned in the same predefined regions.

The same initialization concept but for the entire hand this time is implemented in the HandVU system of Kolsch et al. [23] which employs tracking of multiple features (flock of birds). Figure 5 illustrates the general idea behind systems that employ tracking algorithms: the tracker starts processing when the hand is present in a given location, a necessary condition to compute the values of the features that will be monitored and followed in the succeeding video frames.

Fig. 5 Trackers usually require initialization which consists in presenting the system with the object to be tracked. An initialization event can be triggered whenever the hand is present at some particular location. The features required for the particular tracker are computed when the hand-detected event is triggered.

Malik and Laszlo [31] use location criteria in order to ease the segmentation process when detecting the user's hands for their Visual Touchpad system. The location information is employed indirectly this time and is implemented by using a black colored pad on top of which the hands can be reliably detected due to the high difference in contrast between skin and background colors.

Locations and regions of interest represent powerful event triggers, irrespectively whether they are being used for initialization only or during the actual process of gesture detection and acquisition. They are usually combined with posture events for which predefined hand postures are being required in order for the location events to be confirmed. Two different scenarios can thus be identified for such events that employ location criteria: initialization-only and continuous tracking and monitoring.

2.2 Posture-Based Events

Posture events are triggered whenever a predefined posture is detected. They have been used in order to mark the start and stop timestamps of a motion gesture easing thus the segmentation process considerably [2, 13, 50, 51]. Postures are also commands by themselves and many systems have been developed that use posture recognition solely in order to transmit various commands [13, 24, 26, 49, 50, 60].

Vatavu et al. [51] use a specific hand posture in order to detect when motion commands are entered: recording starts when the finger is pointed and ends when it has been retracted. This gives users better control on how to enter gestures by employing the same principle of click-based interaction. Figure 6 illustrates the segmentation process aided by this specific posture that transforms thus into a motion command detection event. The same click-like events are used by Vatavu and Pentiuc [50] for a TV control application: gesture motions are being recorded and interpreted only when predefined postures are detected (hand open and hand closed have different associated recording actions). The idea is similar to that introduced by Freeman and Weissman [13] but shifted towards a different interaction space that allows privacy, interface sharing and, very important, robustness of the implementation.

Fig. 6 Motion gestures are recorded while the hand is in a predefined posture: the hand closed with the index finger pointed out. The posture acts as an event trigger signaling both the start and stop timestamps for the motion command the user is entering.

A common approach for detecting and tracking human hands in video is to use color-based detectors that learn the distribution of the skin in a given color space. Instead of wearing distinctly colored gloves or rings [54, 64] or using LEDs or any other marker systems [48], skin color detection has proven to be an important intermediate step for face and hands tracking [4, 8, 28]. Other approaches are to track *good* features [47] or sets of such features as in the flock of birds HandVU implementation of Kolsch et al. [24, 26]. The Haar-like features that work so well with face detectors have been equally investigated for detecting various hand postures [25]. Also, more than one camera have been used in order to improve the accuracy of the tracking process [9, 44].

Such advanced and complex tracking algorithms are not always needed for the reliable acquisition of the human hand. For example, Wilson [60] introduced a simple solution for detecting hand gestures in order to simulate cursor functionality and operations on windows. The TAFFI (Thumb and Forefinger) interface avoids complex hand tracking algorithms by simply detecting whether the two fingers are touching and hence forming a closed ellipse-like shape. One and two-handed input is demonstrated for various tasks.

Posture information can be successfully combined with location criteria. For example, the Charade system of Baudel and Beaudouin-Lafon [2] allows a speaker giving a presentation to control a remote computer with free-hand motions while still using gestures in order to communicate with the audience. A numerical data glove measures the amount by which fingers bend as well as the orientation of the hand. Whenever the hand points in a given direction as specified by an active zone, a cursor appears on the screen and starts following the hand motions. Gestures are thus detected when the user's hand is pointing in the active area and actual segmentation of the gesture command is performed using start and stop events defined by specific wrist orientation and finger flexions. Charade is a good example of a system that combines location with posture criteria in order to ease the segmentation process of gestures that can address different functions with different intentions (e.g. controlling the screen and augmenting speech and discourse in front of the audience).

Postures represent thus a powerful cue for segmenting motion gestures as they resemble to some point the well-known and practiced paradigm of click-like interaction. The benefits from the system viewpoint are considerable: motion commands are recorded and processed only when some specific postures are detected. This avoids complex computations for analyzing movements trajectories especially at various scales. The benefits are also important for the human operator as postures create mental models of how the interface works.

2.3 Tap and Touch-Based Events

Touch-based interfaces have been emerging considerably in the last years. The motivation lies in the fact that tapping and touching represent intuitive operations that allow direct contact with the object being selected or manipulated as well as haptic feedback. It is interesting how tap and touch events are strongly connected to interaction intent irrespectively of the technology being used for their detection. Figure 7 illustrates this concept.

Surface computing and interactive tabletops have known a considerable development with the technology used for capturing such touch events being resistive, capacitive or surface wave [40]. For example, Dietz and Leigh [10] describe a technique for creating a touch-sensitive device that allows multiple users to interact simultaneously. Their surface generates modulated electric fields at each location that are capacitively coupled to receivers present in the environment.

Computer vision has also been used for detecting touch and multi-touch events by employing infrared cameras usually placed below a Plexiglas sheet illuminated by an infrared light: when the finger touches the surface, the light becomes frustrated meaning it scatters for the camera to detect it. The concept (frustrated

The Understanding of Meaningful Events in Gesture-Based Interaction

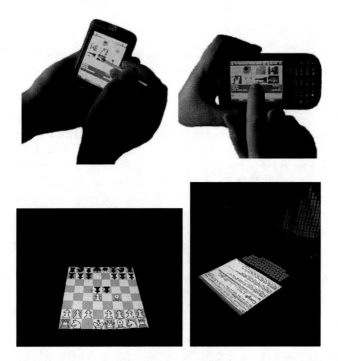

Fig. 7 Tap and touch events specify clearly the intent to interact, both from the system as well as the user's perspective. Top: touch-sensitive mobile devices are becoming more and more popular and affordable. Bottom: an horizontal interactive surface.

total internal reflection) was introduced by Han [15] and proved to be a simple and inexpensive technique for enabling multi-touch sensing at high resolution for interactive surfaces. Computer vision allows such creative and flexible solutions and other approaches have been proposed as well for detecting such touch events. For example, Wilson's system Touchlight [57] uses image processing techniques in order to combine video frames acquired from two infrared cameras placed behind a semi-transparent plane facing the user. By combining the distortion-corrected information from the two video sources, detection of objects that touch or are in a short distance of the surface plane is achieved. When the video camera is placed on top as in the PlayAnywhere system [58], touch events can be detected by comparing the finger position to its shadow.

Tap and touch events specify where and when interaction takes place allowing precise segmentation of gesture motions. Tabletop systems stand thus as an easy-to-use and intuitive technology for interacting with digital content with special techniques and interaction metaphors being proposed [58, 59, 61]. Current applications include browsing photographs, playing videos, listening to music, viewing map locations or ordering menus with the great advantage represented by direct manipulation. Besides the detection of standard touch events, investigations have been carried out in order to support interactions under and above the tabletop [3, 18, 55, 56].

2.4 Custom Events

Next to location, posture and touch, custom events can be specifically defined and implemented in order to augment the gesture acquisition process with regards to accuracy, level of precision or in order to sustain a custom experience. They usually involve additional sensing equipment that needs to be held or worn. For example, Saponas et al. [41] introduced the concept of muscle-computer interfaces that allow gesture-based input by interpreting forearm electromyography with recognition rates of 86% for pinching with one of three fingers [42]. In a much earlier study, Harling and Edwards [16] presented a technique for segmenting the command part of a gesture by using hand tension as the segmentation cue. The investigation started with the observation that the hand is more relaxed as it moves through two boundary (start and stop) postures. Hand tension was defined as the sum of finger tensions while each finger was modeled using elastic springs following Hooke's law.

Various devices that are held or manipulated can be used in order to introduce and take advantage of specific events. One example is the Wii Remote[1] that incorporates accelerometers as well as one infrared camera in order to allow motion sensing. Besides motion analysis, the controller presents additional buttons which may be used in order to specify click-like events. The Wii Remote can be held and manipulated or it can be placed in a fixed location while IR LEDs are controlled instead [30]. Such off-the-shelf devices can even be enhanced for additional sensing. For example, Kry et al. [27] describe a special device constructed on top of a SpaceNavigator by using pressure sensors that detect forces applied by finger tips. This adds posture recognition to position and orientation in order to better control the actions of a virtual hand in the virtual environment.

It needs to be mentioned that a distinct category of events is becoming more and more noticeable: the events used in brain interfaces [36, 43] that rely on various forms of brain activity analysis and for which implementations have been demonstrated even for touch surfaces [63]. A variety of technologies exist for detecting different forms of brain activity and events such as Magnetic Resonance Imagining (MRI), functional MRI (fMRI), Magneto-encephalography (MEG), etc. Out of these, electroencephalograph (EEG) systems provide good recognition accuracy despite being low-cost [29]. They make use of electrodes that are placed on the human scalp in some predefined configuration such as the 10-20 international standard. The acquired EEG signals are then analyzed using pattern recognition techniques.

3 Gestures as Events in the Human-Computer Dialogue

This section explores how the feedback of a correctly detected and recognized gesture is interpreted by users as an important event in the human-computer communication process. As for the human-human dialogue, the gestures of one participant receive corresponding feedback which is interpreted cognitively as well as emotionally. The reason why this happens can be explained by anthropomorphism which, as

---

[1] http://wii.com/

an ancient and natural trait of human beings, applies consequently to computer systems, interfaces and computer-generated output as well. The human operator tends thus to decipher and interpret the response of the system in accordance to previous knowledge gained through social interactions. The deciphering takes place at various levels including affect, behavior and cognition.

The Media equation of Reeves and Naas [39] shows that humans respond to computers as if they were social entities. There are many cases when humans treat computers by attributing them several human characteristics and even adjust their behavior as they would towards other people (with regards to politeness for example). When analyzing this behavior, two different aspects need to be considered: first of all, the developments in anthropomorphic interfaces and, second, the tendency of humans to anthropomorphize the objects they interact with.

With regards to the first aspect, there is a considerable interest in developing interfaces that expose human-like attributes not only in the form of intelligent responses and dialogues but that also incorporate visual human characteristics: virtual characters and actors [5, 6, 11, 14, 53, 62]. Cassell [6] looks at such embodied conversational agent interfaces as:

> "specifically conversational in their behaviors and specifically humanlike in the way they use their bodies in conversation. That is, they may be defined as having the same properties as humans in face-to-face conversation
> ...
> Embodied conversational agents represent a new slant on the argument about whether it is wise to anthropomorphize the interface."

Gratch et al. [14] for example explore the rapports that arise from continuous positive feedback between dialogue partners and show how a virtual character can induce stronger effects than face-to-face communications of human partners. Wagner et al. [53] investigate how realistic such virtual characters need to be in order for the communication and task completion to be effective and engaging. However, even simple virtual actors produce sometimes strong emotions, implication and affect. A simplified and well-known example of a virtual character acting as an interface agent is the Microsoft Office Assistant[2] which was introduced in order to help and assist users with tips to enhance production: Links the cat may seem like a pleasant companion while working (even without the tips) while Rover, the dog, is amusing when searching content in the owner's computer.

The same principles apply even more prominently for human-robot interactions for which the embodiment is present at a much higher level [21, 22, 37]. For example, Kanda et al. [21] explored the possibility that robots could establish relationships with children in order to serve as learning partners, similar to other children. Commercially available smart toys exist such as Sony AIBO[3] which is able to move around, look for toys, play and communicate with the owner. Honda's humanoid robot ASIMO[4] is capable to interpret postures and gestures and move independently

---

[2] http://en.wikipedia.org/wiki/Office_Assistant
[3] http://support.sony-europe.com/aibo/
[4] http://world.honda.com/ASIMO/

in response. Austermann et al. [1] investigated how people interact with both the humanoid ASIMO and the dog-like robot AIBO and found that, although there was no difference in the way commands were sent to the robots, the positive or negative feedback were appropriate to the robot type. For example, the pet AIBO was receiving rewards similarly to a real dog by touch and encouragements while ASIMO received polite personal expressions for gratitude after had accomplished a given task. With great similarity with respect to human-human interactions, a study of Mutlu et. al [35] showed that people perceived ASIMO as being more sociable and more intellectual in a collaboration task than in a competitive one.

Even if the interface is not specifically designed to make use of such human-like elements, the tendency of humans to anthropomorphize is inevitable. This is caused by the emotional implication that humans tend to put into everyday activities which also applies to computers: interacting with a computing system tends to develop a social side as if the machine were another person (see for example the Milo and Kate demo of Lionhead working with the technology of Microsoft Natal project[5]). Another possible explanation would be that anthropomorphism is used as a means to help assimilate new technology. It is also important to note that the feedback of a system will influence humans' responses at the affective, behavioral or cognitive levels.

Other studies have also observed and argued that persons see and interpret feedback by associating and augmenting it with human-like characteristics. For example, Ju and Takayama [20] noticed how people interpret automatic door movements as gestures and found evidence that supports this idea. This goes back to an instinctive reflex of human beings that interpret emotionally the actions they perceive hence the attribution of human characteristics to non-human or even unanimated objects. In an early study, Heider and Simmel [17] showed that people interpret objects that are moving in their visual field in terms of acts of persons. With this respect, emotions, motivations and other human characteristics are being attributed to unanimated objects.

It is therefore important to consider feedback mechanisms that are not only appropriate and adequate in the form they take (visual, audio, haptic, etc.) but also with respect to the emotions they are able to induce. This will transform feedback into a true interaction event perceived as such by the human participant and will lead to an increase in the satisfaction of the overall interaction experience. This is strongly connected to human-human communication where particular gestures are expecting corresponding notification and responses and the same should be expected from human-computer interaction. Appropriate feedback of a correctly detected and recognized gesture can thus become an important communication event.

4 Conclusion

Gesture commands can be detected in video sequences using events that define them in conjunction to location, posture, touch or other application or scenario specific

---

[5] http://www.xbox.com/en-US/live/projectnatal/

criteria. By using such events not only that the amount of processing is reduced which is of extreme importance in vision applications, but the process of gesture segmentation is considerably facilitated. Events can also be intuitive for users which are developing thus a mental model of how the interface works: for example, a valid gesture always starts from one location and ends in another; a given posture specifies the beginning of a command while discarding it ends the recording process; a touch or a tap is clearly associated to the intent to interact. The chapter presented an overview of such events that can be detected in video sequences with specific references to working systems.

While specific events ease both the system as well as the user's input task, it is interesting to note how the system reaction or feedback is interpreted by humans as an important event in the interaction process. Such an event marks a specific interaction point which represents an agreement in communication. The way a system reacts is deciphered and represented by the human operator at multiple levels including affect, behavior and cognition which, in turn, influence the operator's response. This has implications on the way feedback should be reported as well as on the understanding of human-computer interaction by means of general human-human communication.

References

1. Austermann, A., Yamada, S., Funakoshi, K., Nakano, M.: How do users interact with a pet-robot and a humanoid. In: Proceedings of the 28th of the International Conference Extended Abstracts on Human Factors in Computing Systems, CHI EA 2010, Atlanta, Georgia, USA, April 10-15, pp. 3727–3732. ACM, New York (2010)
2. Baudel, T., Beaudouin-Lafon, M.: Charade: remote control of objects using free-hand gestures. Communications of the ACM 36(7), 28–35 (1993)
3. Baudisch, P., Chu, G.: Back-of-device interaction allows creating very small touch devices. In: Proceedings of the 27th International Conference on Human Factors in Computing Systems, CHI 2009, Boston, MA, USA, April 04-09, pp. 1923–1932. ACM, New York (2009)
4. Caetano, T.S., Olabarriaga, S.D., Barone, D.A.C.: Do mixture models in chromaticity space improve skin detection? Pattern Recognition 36(12), 3019–3021 (2003)
5. Cassell, J., Bickmore, T., Billinghurst, M., Campbell, L., Chang, K., Vilhjálmsson, H., Yan, H.: Embodiment in conversational interfaces: Rea. In: Proceedings of the SIGCHI Conference on Human Factors in Computing Systems: the CHI Is the Limit, CHI 1999, Pittsburgh, Pennsylvania, United States, May 15-20, pp. 520–527. ACM, New York (1999)
6. Cassell, J.: Embodied conversational interface agents. ACM Commun. 43(4), 70–78 (2000)
7. Cerlinca, T.I., Pentiuc, S.G., Vatavu, R.D., Cerlinca, M.C.: Hand posture recognition for human-robot interaction. In: Proceedings of the 2007 Workshop on Multimodal Interfaces in Semantic Interaction, WMISI 2007, Nagoya, Japan, November 15, pp. 47–50. ACM, New York (2007)
8. Cho, K.-M., Jang, J.-H., Hong, K.-S.: Adaptive skin color filter. Pattern Recognition 34(5), 1067–1073 (2001)

9. Demirdjian, D., Darrell, T.: 3-D Articulated Pose Tracking for Untethered Deictic Reference. In: Proceedings of International Conference on Multimodal Interfaces, ICMI 2002 (2002)
10. Dietz, P., Leigh, D.: DiamondTouch: a multi-user touch technology. In: Proceedings of the 14th annual ACM symposium on User interface software and technology (UIST 2001), Orlando, Florida, United States, pp. 219–226. ACM Press, New York (2001)
11. Edlund, J., Gustafson, J., Heldner, M., Hjalmarsson, A.: Towards human-like spoken dialogue systems. Speech Commun. 50(8-9), 630–645 (2008)
12. Erol, A., Bebis, G., Nicolescu, M., Boyle, R.D., Twombly, X.: Vision-based hand pose estimation: A review. Computer Vision and Image Understanding 108(1-2), 52–73 (2007)
13. Freeman, W.T., Weissman, C.D.: Television Control by Hand Gestures. In: Proceedings of the 1st International Conference on Automatic Face and Gesture Recognition (1994)
14. Gratch, J., Wang, N., Okhmatovskaia, A., Lamothe, F., Morales, M., Van Der Werf, R.J., Morency, L.: Can virtual humans be more engaging than real ones? In: Jacko, J.A. (ed.) HCI 2007. LNCS, vol. 4552, pp. 286–297. Springer, Heidelberg (2007)
15. Han, J.Y.: Low-cost multi-touch sensing through frustrated total internal reflection. In: Proceedings of the 18th Annual ACM Symposium on User Interface Software and Technology, UIST 2005, Seattle, WA, USA, October 23-26, pp. 115–118. ACM, New York (2005)
16. Harling, P.A., Edwards, A.D.N.: Hand tension as a gesture segmentation cue. In: Progress in Gestural Interaction: Proceedings of Gesture Workshop 1996, pp. 75–87. Springer, Heidelberg (1997)
17. Heider, F., Simmel, M.: An Experimental Study of Apparent Behavior. The American Journal of Psychology 57(2), 243–259 (1944)
18. Hilliges, O., Izadi, S., Wilson, A.D., Hodges, S., Garcia-Mendoza, A., Butz, A.: Interactions in the air: adding further depth to interactive tabletops. In: Proceedings of the 22nd Annual ACM Symposium on User Interface Software and Technology, UIST 2009, Victoria, BC, Canada, October 04-07, pp. 139–148. ACM, New York (2009)
19. Iannizzotto, G., Villari, M., Vita, L.: Hand tracking for human-computer interaction with Graylevel VisualGlove: turning back to the simple way. In: Proceedings of the 2001 Workshop on Perceptive User Interfaces, PUI 2001, Orlando, Florida, November 15-16, vol. 15, pp. 1–7. ACM, New York (2001)
20. Ju, W., Takayama, L.: Approachability: How People Interpret Automatic Door Movement as Gesture. International Journal of Design 3(2) (2009)
21. Kanda, T., Hirano, T., Eaton, D., Ishiguro, H.: Interactive robots as social partners and peer tutors for children: a field trial. Hum.-Comput. Interact. 19(1), 61–84 (2004)
22. Kanda, T., Kamasima, M., Imai, M., Ono, T., Sakamoto, D., Ishiguro, H., Anzai, Y.: A humanoid robot that pretends to listen to route guidance from a human. Auton. Robots 22(1), 87–100 (2007)
23. Kolsch, M., Turk, M., Hollerer, T.: Vision-Based Interfaces for Mobility. In: Proceedings of the International Conference on Mobile and Ubiquitous Systems, MobiQuitous 2004 (2004)
24. Kolsch, M., Hollerer, T., DiVerdi, S.: HandVu: A New Machine Vision Library for Hand Tracking and Gesture Recognition, demo at ISWC/ISMAR (2004)
25. Kolsch, M., Turk, M.: Robust Hand Detection. In: Proceedings of the IEEE International Conference on Automatic Face and Gesture Recognition (2004)
26. Kolsch, M., Turk, M.: Hand tracking with Flocks of Features. In: Proceedings of the IEEE Conference on Computer Vision and Pattern Recognition (2005)

27. Kry, P.G., Pihuit, A., Bernhardt, A., Cani, M.: HandNavigator: hands-on interaction for desktop virtual reality. In: Proceedings of the 2008 ACM Symposium on Virtual Reality Software and Technology, VRST 2008, Bordeaux, France, October 27-29, pp. 53–60. ACM, New York (2008)
28. Lee, J.Y., Yoo, S.I.: An elliptical boundary model for skin color detection. In: Proceedings of the Int. Conf. on Imaging Science, Systems and Technology, Las Vegas, USA (2002)
29. Lee, J.C., Tan, D.S.: Using a low-cost electroencephalograph for task classification in HCI research. In: Proceedings of the 19th Annual ACM Symposium on User interface Software and Technology, UIST 2006, Montreux, Switzerland, October 15-18, pp. 81–90. ACM, New York (2006)
30. Lee, J.C.: Hacking the Nintendo Wii Remote. IEEE Pervasive Computing 7(3), 39–45 (2008)
31. Malik, S., Laszlo, J.: Visual touchpad: a two-handed gestural input device. In: Proceedings of the 6th International Conference on Multimodal Interfaces, State College, PA, USA, October 13-15, pp. 289–296. ACM, New York (2004)
32. Marcel, S.: Hand posture recognition in a body-face centered space. In: CHI 1999 Extended Abstracts on Human Factors in Computing Systems, Pittsburgh, Pennsylvania, May 15-20, pp. 302–303. ACM, New York (1999)
33. Moeslund, T.B., Granum, E.: A Survey of Computer Vision-Based Human Motion Capture. Computer Vision and Image Understanding 81(3), 231–268 (2001)
34. Moeslund, T.B., Hilton, A., Kruger, V.: A survey of advances in vision-based human motion capture and analysis. Computer Vision and Image Understanding, Special Issue on Modeling People: Vision-based understanding of a person's shape, appearance, movement and behaviour 104(2-3), 90–126 (2006)
35. Mutlu, B., Osman, S., Forlizzi, J., Hodgins, J., Kiesler, S.: Perceptions of ASIMO: an exploration on co-operation and competition with humans and humanoid robots. In: Proceedings of the 1st ACM SIGCHI/SIGART Conference on Human-Robot Interaction, HRI 2006, Salt Lake City, Utah, USA, March 02-03, pp. 351–352. ACM, New York (2006)
36. Nijholt, A., Tan, D., Allison, B., Milan, J.d.R., Graimann, B.: Brain-computer interfaces for HCI and games. In: CHI 2008 Extended Abstracts on Human Factors in Computing Systems, pp. 3925–3928. ACM, New York (2008)
37. Okuno, Y., Kanda, T., Imai, M., Ishiguro, H., Hagita, N.: Providing route directions: design of robot's utterance, gesture, and timing. In: Proceedings of the 4th ACM/IEEE International Conference on Human Robot Interaction, HRI 2009, La Jolla, California, USA, March 09-13, pp. 53–60. ACM, New York (2009)
38. Poppe, R.: Vision-based human motion analysis: An overview. Computer Vision and Image Understanding 108(1-2), 4–18 (2007)
39. Reeves, B., Nass, C.: The Media Equation: how People Treat Computers, Television, and New Media Like Real People and Places. Cambridge University Press, Cambridge (1996)
40. Saffer, D.: Designing Gestural Interfaces. O'Reilly Media Inc., Sebastopol (2009)
41. Saponas, T.S., Tan, D.S., Morris, D., Balakrishnan, R., Turner, J., Landay, J.A.: Enabling always-available input with muscle-computer interfaces. In: Proceedings of the 22nd Annual ACM Symposium on User Interface Software and Technology, UIST 2009, Victoria, BC, Canada, October 04-07, pp. 167–176. ACM, New York (2009)

42. Saponas, T.S., Tan, D.S., Morris, D., Turner, J., Landay, J.A.: Making muscle-computer interfaces more practical. In: Proceedings of the 28th International Conference on Human Factors in Computing Systems, CHI 2010, Atlanta, Georgia, USA, April 10-15, pp. 851–854. ACM, New York (2010)
43. Sauvan, J., Lcuyer, A., Lotte, F., Casiez, G.: A performance model of selection techniques for P300-based brain-computer interfaces. In: Proceedings of CHI 2009, pp. 2205–2208. ACM, New York (2009)
44. Schlattman, M., Klein, R.: Simultaneous 4 gestures 6 DOF real-time two-hand tracking without any markers. In: Spencer, S.N. (ed.) Proceedings of the 2007 ACM Symposium on Virtual Reality Software and Technology, VRST 2007, Newport Beach, California, November 05-07, pp. 39–42. ACM, New York (2007)
45. Jaimes, A., Sebe, N.: Multimodal human-computer interaction: A survey. Computer Vision and Image Understanding 108(1-2), 116–134 (2007)
46. Sharma, N.K., Kurian, G.: Language, Thought and Communication. In: Krishnan, L., Patnaik, B.N., Sharma, N.K. (eds.) Aspects of human communication. Mittal Publications (1989)
47. Shi, J., Tomasi, C.: Good features to track. In: Proceedings of the IEEE Conference on Computer Vision and Pattern Recognition (1994)
48. Sturman, D.J., Zeltzer, D.: A Survey of Glove-based Input. IEEE Computer Graphics and Applications 14(1), 30–39 (1994)
49. Vatavu, R.D., Pentiuc, S.G., Chaillou, C., Grisoni, L., Degrande, S.: Visual Recognition of Hand Postures for Interacting with Virtual Environments. In: Proceedings of the 8th International Conference on Development and Application Systems - DAS 2006, Suceava, Romania, pp. 477–482 (2006)
50. Vatavu, R.D., Pentiuc, S.G.: Interactive Coffee Tables: Interfacing TV within an Intuitive, Fun and Shared Experience. In: Tscheligi, M., Obrist, M., Lugmayr, A. (eds.) EuroITV 2008. LNCS, vol. 5066, pp. 183–187. Springer, Heidelberg (2008)
51. Vatavu, R.D., Grisoni, L., Pentiuc, S.G.: Gesture Recognition Based on Elastic Deformation Energies. In: Sales Dias, M., Gibet, S., Wanderley, M.M., Bastos, R. (eds.) GW 2007. LNCS (LNAI), vol. 5085, pp. 1–12. Springer, Heidelberg (2009)
52. Viola, P., Jones, M.: Rapid object detection using a boosted cascade of simple features. In: Proceedings of the IEEE Conference on Computer Vision and Pattern Recognition, pp. 511–518 (2001)
53. Wagner, D., Billinghurst, M., Schmalstieg, D.: How real should virtual characters be? In: Proceedings of the 2006 ACM SIGCHI international Conference on Advances in Computer Entertainment Technology, ACE 2006, Hollywood, California, June 14-16, vol. 266, p. 57. ACM, New York (2006)
54. Wang, R.Y., Popovic, J.: Real-time hand-tracking with a color glove. In: Hoppe, H. (ed.) ACM SIGGRAPH 2009 Papers, SIGGRAPH 2009, New Orleans, Louisiana, August 03-07, pp. 1–8. ACM, New York (2009)
55. Wigdor, D., Leigh, D., Forlines, C., Shipman, S., Barnwell, J., Balakrishnan, R., Shen, C.: Under the table interaction. In: Proceedings of the 19th Annual ACM Symposium on User interface Software and Technology, UIST 2006, Montreux, Switzerland, October 15-18, pp. 259–268. ACM, New York (2006)
56. Wigdor, D., Forlines, C., Baudisch, P., Barnwell, J., Shen, C.: Lucid touch: a see-through mobile device. In: Proceedings of the 20th Annual ACM Symposium on User Interface Software and Technology, UIST 2007, Newport, Rhode Island, USA, October 07-10, pp. 269–278. ACM, New York (2007)

57. Wilson, A.D.: TouchLight: an imaging touch screen and display for gesture-based interaction. In: Proceedings of the 6th International Conference on Multimodal Interfaces, ICMI 2004, State College, PA, USA, October 13-15, pp. 69–76. ACM, New York (2004)
58. Wilson, A.D.: PlayAnywhere: a compact interactive tabletop projection-vision system. In: Proceedings of the 18th Annual ACM Symposium on User Interface Software and Technology, UIST 2005, Seattle, WA, USA, October 23-26, pp. 83–92. ACM, New York (2005)
59. Wilson, A., Robbins, D.C.: Playtogether: Playing games across multiple interactive tabletops. In: IUI Workshop on Tangible Play: Research and Design for Tangible and Tabletop Games (2006)
60. Wilson, A.D.: Robust computer vision-based detection of pinching for one and two-handed gesture input. In: Proceedings of the 19th Annual ACM Symposium on User Interface Software and Technology (UIST 2006), Montreux, Switzerland, pp. 255–258. ACM Press, New York (2006)
61. Wu, M., Balakrishnan, R.: Multi-finger and whole hand gestural interaction techniques for multi-user tabletop displays. In: Proceedings of the 16th Annual ACM Symposium on User Interface Software and Technology, UIST 2003, Vancouver, Canada, November 02-05, pp. 193–202. ACM, New York (2003)
62. Xiao, J.: Understanding the use and utility of anthropomorphic interface agents. In: CHI 2001 Extended Abstracts on Human Factors in Computing Systems, CHI 2001, Seattle, Washington, March 31-April 05, pp. 409–410. ACM, New York (2001)
63. Yuksel, B.F., Donnerer, M., Tompkin, J., Steed, A.: A novel brain-computer interface using a multi-touch surface. In: Proceedings of the 28th International Conference on Human Factors in Computing Systems, CHI 2010, Atlanta, Georgia, USA, April 10-15, pp. 855–858. ACM, New York (2010)
64. Zhang, L.-G., Chen, Y., Fang, G., Chen, X., Gao, W.: A vision-based sign language recognition system using tied-mixture density HMM. In: ICMI 2004: Proceedings of the 6th International Conference on Multimodal Interfaces, State College, PA, USA, pp. 198–204. ACM Press, New York (2004)

Apply GPCA to Motion Segmentation

Hongchuan Yu and Jian J. Zhang

Abstract. In this paper, we present a motion segmentation approach based on the subspace segmentation technique, the generalized PCA. By incorporating the cues from the neighborhood of intensity edges of images, motion segmentation is solved under an algebra framework. Our main contribution is to propose a post-processing procedure, which can detect the boundaries of motion layers and further determine the layer ordering. Test results on real imagery have confirmed the validity of our method.

1 Introduction

An important problem in computer vision is to segment moving objects of a scene from a video source, and partly recover the structure or motion information, such as foreground and background. With widespread demands on video processing, motion segmentation has found many direct applications. Video surveillance systems seek to automatically identify people, objects, or activities of interest in a variety of environments with a set of stationary cameras. Motion segmentation can provide low level motion detection and region tracking cues. Another relatively new application is markerless motion capture for computer animation. It aims to estimate the human body configuration and pose in the real world from a video by locating the joint positions over time and extracting the articulated structure.

Motion segmentation is expected to partly recover the structure and motion information of moving objects from a mutually occluded scene. This includes the following main tasks, (1) labeling the regions of a motion layer segmentation, i.e. pixels are assigned to several motion layers; (2) finding their motion, e.g. each layer has its own smooth flow field while discontinuities occur between layers; (3) determining the layer ordering, as the different layers might occlude each other. But motion segmentation is not equivalent to object tracking. Roughly speaking, object

---

Hongchuan Yu · Jian J. Zhang
NCCA, Bournemouth University, Poole, U.K.
e-mail: {hyu,jzhang}@bournemouth.ac.uk

tracking is to track the segmented objects over an image sequence, although the extension of the rigidity constraint to multiple frames is nontrivial. Motion segmentation aims at the motion layers of a scene rather than the moving objects. For example, if a moving object contains multiple motions at a moment, it may be divided into several motion layers. When these motion layers share the same motion, they could be merged into a single layer. Hence, motion segmentation usually uses the information from a few successive frames. In contrast, object tracking focuses on a moving object in a scene. It utilizes the information from an image sequence. Motion segmentation plays a role of fundamental module in motion analysis and tracking. [14] presented a subspace segmentation method to estimate the motion models of the motion layers based on two successive frames. Built on this subspace segmentation method, this paper will further aim at two other basic problems of motion segmentation, i.e. the detection of motion layer boundaries and depth ordering based on two successive frames. The basic idea is to refine a global segmentation to solve these two problems. We first address this subspace segmentation approach for motion model estimation. We then incorporate it with the intensity edge information into a post-processing procedure, which refines the layer boundaries and infers the layer order between two successive frames. These two procedures form a complete algorithm for motion segmentation. Our specific contributions in this paper include 1) the Polysegment algorithm (a special case of the generalized PCA [12]) is employed to detect the layer boundaries in our post-processing procedure, and 2) the cues from the intensity edges of images are utilized in the detection of the layer boundaries and depth ordering.

1.1 Previous Works

Although motion segmentation has long been an active area of research, many issues remain open in computer vision, such as the estimation of multiple motion models [1,2], layered motion descriptions [3,4], occlusion detection and depth ordering [5-7].

Most popular approaches to motion segmentation revolve around parsing the optical flow field in an image sequence. Because of the well-known aperture problem, the motion vector from optical flow computation can only be determined in the direction of the local intensity gradient. For the sake of completeness of optical flow field, it is assumed that the motion is locally smooth. Obviously, depth discontinuities and multiple independently moving objects usually result in discontinuities of the optical flow. The usual approaches are to parameterize the optical flow field and fit a different model (e.g. 2D affine model) to each moving object, such as the layered representation of the motion field [3]. The challenges of the optical flow-based techniques involve identifying motion layers (or pixel grouping), detecting layer boundaries, and depth ordering. Previous research can mostly be grouped into two categories. The first category is to determine all of the motion models simultaneously. This can be achieved by parameterising the motions and segmentation, and using sophisticated statistical techniques to predict the most probable solution. For example, Smith et al. in [6] presented a layered motion segmentation approach under a Bayesian framework by tracking edges between frames. In the implementation of their proposed scheme, the region edge labels were not directly applied to the Bayesian model. They were implicitly determined by the foreground-background orders of the motion layers and the motion layer labels for each region. Kumar et al. in [8] presented the learning approach of a generative layered representation of a scene for motion segmentation. In order to get the initial estimates of model, they utilized the loopy belief propagation, and further refined the initial estimate by using $\alpha\beta$-swap and α-expansion algorithms. The large number of undetermined parameters in their Bayesian models leads to the difficult tracking problem in a high dimensional parameter space. The second category is the dominant motion approach [9-11]. A single motion is first fitted to all pixels, and then to test for pixels that agree with that motion. This process can be repeated recursively on the outlier pixels to provide a full set of layers [10]. The central problem faced by this kind of approaches is that it is extremely difficult to determine the occluded edges of the moving regions (or motion layers). Furthermore, this problem can result in the failure of depth ordering of motion layers. However, analytically reasoning such complex cases is impractical. The main reasons are three fold. First, the smoothing required by the optical flow algorithms makes it difficult to localize the layer boundaries. Second, the optical flow field is usually parameterized by some 2D motion models (e.g. 2D affine), which is the first order approximation of the perspective model. It is unreliable to apply a 2D model to the boundaries of moving regions. Third, pixels in a neighborhood of the boundaries are in the areas of high intensity gradient. Slight errors or image noise can result in pixels of a very different intensity, even under the correct motion estimate [6]. In this paper, we will simplify the problem of motion segmentation based on an algebraic framework. We will first obtain a rough global segmentation and then refine it afterwards.

Our work is partially inspired by the subspace segmentation approach to motion model estimation proposed in [14]. This approach can provide a non-iterative and global estimation of motion layer segmentation. But it is incomplete, since the depth ordering and the detection of layer boundaries are ignored. In this paper we provide a complete solution by developing a novel post-processing procedure using the intensity structures of edges for the detection of (1) motion layer boundaries and (2) the layer order.

In the remainder of this paper, we first briefly review the subspace segmentation approach to motion model estimation [14] in section 2. In section 3, a post-processing procedure is presented for the detection of the layer boundaries and depth ordering. The experimental results and analysis are given in section 4. Our conclusion and future work are given in section 5.

2 Motion Segmentation by GPCA-PDA

The core of our proposed motion segmentation approach is the scheme of segmenting hyperplanes in R^K, which is called the generalized PCA (GPCA) in [12]. Applying the GPCA method to motion model estimation has been proposed in

[14]. But the resulting motion model estimation can only yield coarse motion segmentation, i.e. the boundary of the motion layers is very blurry. Our basic idea is to further refine the boundary of the resulting motion layers by a post-processing procedure. Before introducing our post-processing procedure, we firstly review the motion model estimation approach in [14] briefly. The two used algorithms, GPCA-PDA Alg. and Polysegment Alg., can be found in [12]. (We also briefly introduce these two algorithms in Appendix.)

The first problem to motion segmentation is to obtain the layered motion models corresponding to independently moving regions in a scene, (i.e. layer segmentation). We address an algebra approach in terms of a known optical flow field which has been presented in [14]. Its distinct advantage over the other approaches is that it can determine all motion layers simultaneously.

Given N measurements of the optical flow $\{(u,v)_i\}_{i=1}^{N}$ at the N pixels $\{(x_1,x_2)_i\}_{i=1}^{N}$, we can describe them through a affine motion as follows,

$$\begin{cases} a_{11}x_1 + a_{12}x_2 + a_{13} - u = 0 \\ a_{21}x_1 + a_{22}x_2 + a_{23} - v = 0 \end{cases}.$$

In terms of the hyperplane representation in the Appendix, the solution to the multiple independent affine models can be rephrased as follows. Let $x = (x_1, x_2, 1, u, v)^T \in R^5$ and hyperplane S_i be spanned by the basis of $b_1 = (a_{11}, a_{12}, a_{13}, a_{14}, 0)^T$ and $b_2 = (a_{21}, a_{22}, a_{23}, 0, a_{24})^T$. We need to segment a mixture of the hyperplanes of dimension $d = 3$ in R^5, which is expressed as,

$$S_i = \left\{ x \in R^5 : (b_1, b_2)_i^T x = 0 \right\}.$$

The original equations of optical flow have finished the projection from $x \in R^5$ to two individual subspaces of R^4 in a natural way, i.e. each new hyperplane in R^4 can be expressed as,

$$(a_{11}, a_{12}, a_{13}, a_{14}) \cdot (x_1, x_2, x_3, x_4) = 0.$$

Applying the scheme of Eq.(A1-A4) in Appendix can yield the desired basis $B^{(i)} = (b_1, b_2)_i$ for each hyperplane S_i in R^3.

Up to now, one can obtain the initial estimation of all of the motion layers simultaneously. This is insufficient for motion segmentation, since we also need to determine the layer boundaries and the occlusion relationship. Beside that, it can be observed that each segmented layer contains some small and isolated spurious regions, and the resulting layer boundaries wander around the real ones. This makes the detection of the layer boundaries difficult. The occluded regions take place in the neighborhood of the layer boundaries. If the occluding edges can be determined correctly, the occluded regions can be segmented correctly. Furthermore, the resulting motion layers can also be linked to the occluded regions in terms of the occluding edges for the depth ordering. Hence, it is a crucial step to determine the occluding edges. Our development is based on the following observations (1) the intensity edges include the boundaries of motion layers; (2) the layer boundaries are not always the occluding edges; (3) determining the occluding edges and inferring the occlusion relationship can be fulfilled by testing the neighborhood of edges.

We will introduce the intensity edges of images into the potential occlusion areas for the detection of occluding edges in the next section.

3 Post-Processing Procedure

Let us consider a single viewpoint. The central problem is to detect the occluding edges, because the erroneous edge labeling can cause incorrect depth ordering. Most of the techniques considered so far employed only the motion field information for motion segmentation. For each frame, all edges, including edges of motion layers and textured edges of objects, are presented in the image intensity structure, which can provide the wealth of additional information to motion estimation. Due to their extreme length, a number of measurements might be taken along (or around) them. This leads to a more accurate estimation of motion.

Recent applications have motivated a renewal of motion segmentation by tracking edges [6,17]. Ogale et al. [7] classified the occlusions into three classes. In order to deduce the ordinal depth, they had to fill the occluded regions. This is to implicitly approximate the occluding edges by filling the neighborhood of the layer boundaries. [6] provides three fundamental assumptions of the relationship between regions and edges to identify the edges of moving regions. We add an extra assumption (i.e. the 4th below) and emphasize these four assumptions as follows.

(1) As an object moves all of the edges associated with that object move, with a motion, which may be approximately described by some motion model.
(2) The motions are layered, i.e. one motion takes place completely in front of another, and the layers are strictly ordered. Typically the layer farthest from the camera is referred to as the background, with nearer foreground layers in front of this.
(3) An arbitrary segmented image region only belongs to one motion model, and hence any occluding boundary is visible as a region edge in the image.
(4) For each frame, the intensity edges involve the edges of motion layers.

An important conclusion from these four assumptions is that the layer ordering can be uniquely determined if the layer of each moving region is known and the occluding edges are known. [7] presented the relationship of motion layers and occluded regions, and further emphasized that the motion layer involving the occluded region must be behind another one. Even when the layers of motion regions are known, ambiguities may still be presented in the layer boundary labeling, as shown in Fig.1. In Fig.1a, due to the occluded region C, we can infer the occlusion relationship between the motion regions A and B, while, in Fig.1b, we

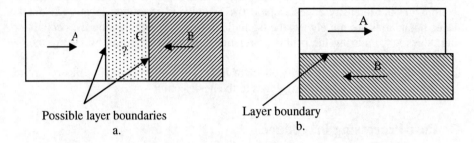

Possible layer boundaries
a.

Layer boundary
b.

Fig. 1 Illustration of moving regions (*A,B*) and occluded region (*C*). (a) The probable layer boundaries are determined by extending the moving region to the occluded region; (b) There is no occlusion region between layers *A* and *B*.

cannot find out the layer order according to the distinct edges of the motion layers. The layer boundaries are not the same as the occluding edges. The layer boundaries involve the occluding edges, but the layer boundaries are not always the occluding edges. It is infeasible to infer the layer order only by the layer boundaries. We can therefore conclude that the occlusion relationship hides behind the occluded regions, and identifying the occluding edges can reveal the occlusion relationship. The optical flow computation can usually identify the coarse occlusion regions as a by-product [15], which will be adopted in this paper.

The subspace segmentation approach described in section 2 is carried out on a given optical flow field instead of the image intensities. Due to the errors from the optical field (e.g. aperture problem etc.), each resulting motion layer contains two kinds of artifacts: (1) small isolated regions with texture and (2) dark holes over the image plane. It can be observed that a single hole in the middle of a foreground layer runs through to the background layer. Similar problems also exist in the occluded regions. Moreover, the resulting boundaries of motion layers and their neighborhood are, in general, highly unreliable areas. Therefore, the segmentation by the subspace segmentation method and the occluded regions detected by the optical flow computation cannot offer a valid solution to the above two problems.

Consider the neighborhood of the layer boundaries. It can be observed that the occluded regions are involved in the neighborhood of the layer boundaries as shown in Fig.1a. The edges' neighborhood contains the wealthy intensity structures of image. This can provide us sufficient cues to find the layer boundaries and occluding edges. We rephrase the problem of layer edge detection and depth ordering, and present our post-processing procedure as follows.

The motion models of the layers are determined by the subspace segmentation approach described in section 2, while the layer boundaries and the layers of the occluded regions are undetermined. The problem we face here is how to determine the layer boundaries and infer the occlusion relationships. In order to do that, we will consider the intensity structures of each frame, the relevant occlusion region map (obtained by [15]) and the relevant boundary map of the initial motion layers (obtained by the subspace segmentation approach). Let us denote intensity edge map as M_I, occlusion region map as M_O and layer edge map as M_L thereafter. The motion of intensity edges dominates that of their neighborhood. It is straightforward to utilize the intensity structures of the neighborhood of the edges for detecting the layer boundaries and inferring the occlusion relationship. The proposed post-processing procedure given below is performed over two successive frames, but evidence could be accumulated over an image sequence for a more robust segmentation.

Construct Pending Areas

For each frame, we first determine some pending areas, which should involve all potential layer boundaries. Then, the detection of layer boundaries is carried out on the resulting pending areas accordingly. To this end, we place a set of windows w of size $n \times n$ along the edges of M_L. These small windows might be overlapped to each other. Usually each window w_i is determined by the M_O and M_L without a fixed size, i.e. it is expected to be so large that the resulting set of windows can cover the occlusion regions M_O and layer edge map M_L on the current frame. In our experiments, the minimal size n of w_i is set to 10 pixels.

Match Scores

Consider the resulting pending areas $W = \bigcup_i w_i$, which contains many intensity edges $l \subset M_I$. The potential layer boundaries are involved in M_I in terms of the assumption (4). Thus, for each window w_i, we can compute the profile of every point p, which is defined as a vector $pf(p)$ by sampling the intensity derivative in the positive and negative directions of the intensity gradient at p. This is illustrated in Fig.2. The point profile is then normalized as,

$$pf(p) = \frac{pf(p)}{\sum_j |pf_j(p)|}. \qquad (1)$$

According to the optical flow field, one can get a pair of corresponding points p and p' respectively on two successive frames. The match score is taken as the residual error of their profiles as follows,

$$e_0(p) = Exp\left\{-\frac{\|pf^{(i)}(p) - pf^{(i+1)}(p')\|^2}{\sigma^2}\right\},$$

where σ is a reference distance and is determined empirically. When the point p is far away from layer boundaries, $e_0(p)$ should approach one, i.e. the neighborhood of p obeys a single motion. Otherwise its neighborhood contains multiple motions. Furthermore, we can obtain other two match scores respectively along either side profile of the current point p, denoted as $e_1(p), e_2(p)$. If point p belongs to a layer boundary, one of these two scores should approach one while the other should approach zero, otherwise both of them should approach one.

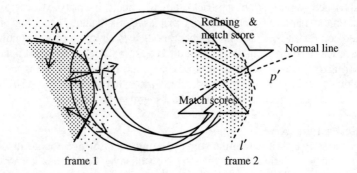

Fig. 2 Illustration of point profiles and refining the matching point. Refining procedure is carried out as 1D search along the normal line on the frame 2.

Matching

Because of the aperture problem, the motion of edges can only be determined in the direction normal to the edge. This means that the corresponding point p' of the next frame (i+1) lies on the normal line, which is the normal at point p on the current frame (i). This is useful as it restricts matching isointensity contour on the frame (i+1) along the edge normal. In order to enhance the intensity edge matching, we add a new match score that is the residual error $e_0^\perp(p)$ of the profiles of p and p' along the edge tangent line, which is shown in Fig.2. Refining the matching point p' on the next frame (i+1) is thus implemented as a 1D search based on the match score of $\left(e_0(p)+e_0^\perp(p)\right)$ along the direction of point p's gradient (i.e. the normal line) instead of point p''s gradient, which is also illustrated in Fig.2. After that, one can re-compute the match scores e_0, e_1, e_2 of the points p and p' in terms of their individual intensity gradients rather than the normal line.

Segmentation by Polysegment Alg.

Based on the match scores e_0, e_1, e_2 in the pending areas W, we apply the Polysegment algorithm as described in the Appendix respectively to the match scores of e_0, e_1, e_2 for the layer edge detection. There are two groups here, one is the group of layer edge points and the other is that of non-layer edge points. For each match score, we can thus get two cluster centers $\mu^{(i)} = \left(\mu_1^{(i)}, \mu_2^{(i)}\right), i = 0,1,2$. Moreover, to the points $p \in W$, there are eight cluster centers. The layer edge points should cluster around the two centers of $cent_1 = \left(\min \mu^{(0)}, \min \mu^{(1)}, \max \mu^{(2)}\right)$ and $cent_2 = \left(\min \mu^{(0)}, \max \mu^{(1)}, \min \mu^{(2)}\right)$. The segmentation of W is obtained as follows,

$$i = \arg\min_{j=1,\dots,8} \|e(p) - cent_j\|^2, \qquad (2)$$

where $e(p) = (e_0, e_1, e_2)(p)$. On this basis, the points of W can be classified into two groups, layer edge points and non-layer edge points.

Region Merging

For the group of non-layer edge points, one can merge most of small spurious regions in a big motion layer, i.e. merging small regions with a motion layer by comparing their areas with their individual neighbors'. This can lead to the connected layer. But it can be observed that the detected layer boundaries usually have discontinuities with the group of layer edge points, i.e. a set of layer edge segments. This is due to the fact that some layer edge points are incorrectly classified into the group of non-layer edge points. Based on the areas of the segmented layer regions, it is impossible to make a correct decision of region merging when these small regions may contain layer edge segments. This is because the layer edge segments indicate that the both sides should respectively occupy different motion layers and could not be merged into a single layer at anytime. These regions are thus left as the undetermined regions temporarily.

On the other hand, a connected layer has a continuous boundary M_L. These layer edge segments only prune the region of the layer, but do not form new closed regions within the layer. In our experiments, we simply replace some parts of M_L with the new layer edge segments according to the nearest neighbor criterion. Then, the area comparison strategy is employed to those undetermined regions nearby the layer boundary for region merging. A layer edge segment separates one region into two motion layers. When merging two or more undetermined regions which share a layer edge segment, the merging procedure should be terminated.

Depth Ordering

After region merging, one can obtain the desired boundaries of motion layers M_L. If the occluded regions belong to a motion layer, this layer must be behind another one. Our problem can now be rephrased as HOW to assign the occluded regions to the known motion layers.

With the occlusion region map M_O, one can first determine which points of the layer boundaries belong to the occluding edges, since the layer boundaries involve the occluding edges. The worst case is that the points of the occluding edges are not within M_O. But both side profiles of these points should overlap with M_O at that moment. On this basis, one can determine the points of the occluding edges by checking if they are within M_O or their profiles overlap with M_O. Since some points of layer boundaries may not belong to the occluding edges, such as in Fig.1b, the depth ordering can only carry out on the detected occluding edges. Then, for the points of the occluding edges, one can extend their profiles in the direction of the intensity gradient to the known motion layers for their profile labeling, i.e. inferring which layers both sides of the occluding edges respectively belong to.

Furthermore, inferring the occlusion relationship can be fulfilled by comparing the match scores $e_1(p), e_2(p)$ of each point p of the occluding edges. This is because an occluded region only shares the same motion layer with one of the profiles of an occluding edge. The smaller match score corresponds to the real occlusion region. This implies that one side of an occluding edge with the smaller match score is behind the other side, since it involves the occlusion region. In terms of the profile labeling of the occluding edge points, we can therefore find ordinal depth.

The **Post-processing procedure** is summarized as follows:

(1) Extracting the pending areas W on each frame;
(2) Refining the corresponding points p' on the next frame (i+1), and then recompute the match scores $e_0(p), e_1(p), e_2(p)$;
(3) Applying the polysegment algorithm to the match scores of W for detecting the points of layer boundaries;
(4) Merging the spurious regions for the continuous boundaries of the motion layers;
(5) Determining the occluding edges in terms of M_O;
(6) Extending the profiles of the occluding edge points to the known motion layers for the profile labeling;
(7) Comparing the match scores $e_1(p), e_2(p)$ of the occluding edge points p for depth ordering, i.e. $\min\{e_1(p), e_2(p)\}$ corresponds to the occluded region.

This post-processing procedure and the subspace segmentation approach described in section 2 constitute a complete algorithm of motion segmentation. Note that in our algorithm, the estimation of all the motion models in a scene is undertaken at the first procedure (i.e. subspace segmentation method), and the detection of the layer boundaries and depth ordering are carried out at the second procedure (i.e. post-processing procedure). This is different from the previous approaches. Usually the motion model estimation was mixed with the later processing. This makes the algorithms complicated and the implementation difficult.

4 Experiments and Analysis

Our algorithm was tested on several image sequences. In this section, the three results of the 'flower garden', 'Susie calling' and 'table tennis' are presented. All programs have been implemented on the MatLab platform using a publicly available package—GPCA-PDA [16] and the optical flow code in [15]. All the image sequences used in our experiments are available at [18].

Flower Garden

In this experiment, we applied our motion segmentation approach to the flower garden sequence of resolution 175×120 pixels. The tree trunk in front of a garden

Apply GPCA to Motion Segmentation

is taken by a camera undergoing translation from the left to the right. Our goal is to determine the boundaries of the motion layers, and find the layer order over two successive frames. Our approach found out two motion layers, the tree trunk and garden background.

Fig.3 gives the segmentation of the affine model using the subspace segmentation approach described in section 2. It can be noted that the occlusion regions from the optical flow fields are crude, and contain many spurious small regions. The red arrows illustrate the possible occlusion regions in the successive frame 1 and 2. We show the results of motion segmentation of frame 1 in Fig.3(1-3). Note that the occlusion regions in Fig.3(2-3), which are not the layer boundaries, are the interim areas between the foreground and background. It is impossible to determine the depth ordering using the obtained motion layers before determining the layers of the occluded regions. In addition, the resulting layer boundaries are also unreliable. Similar to the occluded regions, there are many small and isolated spurious regions on the obtained layers. We need to refine the layer boundaries and find out the layer order.

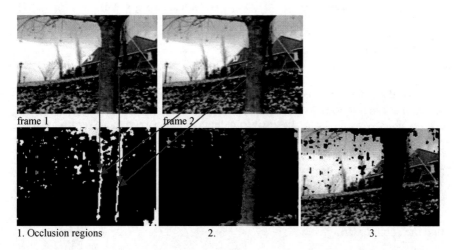

Fig. 3 Motion segmentation results by the subspace segmentation approach only. 1) The occlusion regions from the optical flow field between frame 1 and 2; 2) and 3) are the segmentation results only using the subspace segmentation approach described in section 2.

Fig.4 gives the segmentation results of the subspace segmentation approach followed by the post-processing procedure described in section 3. The boundaries of motion layers can go across the occluded regions and converge to the desired locations. But we can also note that a patch of ground is classified as the foreground as shown in Fig.4(1). This is due to the fact that the motion variance of this patch between the successive frames 1 and 2 is close to that of the "tree", away from the background. If going through over multiple frames, the motion of this patch should be distinguished from the "tree", since the motion of the ground is prone to be modeled by a single affine model. The intensity edge map of frame 1

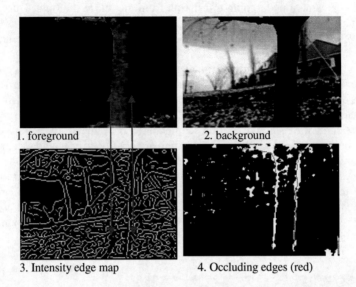

Fig. 4 Refined and layered motion segmentation. 1) and 2) are the segmentation results; 3) The intensity edges of the image; 4) Occluded region and occluding edges.

is obtained by the Canny edge detector, and also shown in Fig.4(3). It can be observed that the boundaries of motion layers are involved in the intensity edge map, e.g. the red arrows illustrate the corresponding edges between the layer boundaries and the intensity edges of the image. Moreover, we also show the occlusion edges (red) in Fig.4(4). Partial occluding edges are not involved in the initial occlusion regions. But, the profiles of these edge points overlap with the occlusion regions. These points can thus be joined with the occluding edges. Additionally, it can be noted that the layer boundaries involve the occluding edges, but the layer boundaries are not always the occluding edges. Locating the occluding edges can help us find the depth ordering.

Susie Calling

This sequence presents a hand holding a phone while the head is rising slightly. The image resolution is 170×120 pixels. It can be observed that the region of the phone is enwrapped by the head region. Our segmentation approach aims at separating the phone region from the head region. The segmentation results are shown in Fig.5. The region of phone is in front of the regions of head and background. The background region is behind the head region.

Due to the rich texture of the hair, the segmented head region contains many small holes, particularly in the hair area. It is difficult to determine the boundaries of the hair. For example, in Fig.5(2), a patch of hair image is incorrectly classified into the group of the phone region. The post-processing could not merge this patch into the head region either, as shown in Fig.5(4-6). This is because the detected layer edge segments goes through the occluded regions as shown in Fig.5(7).

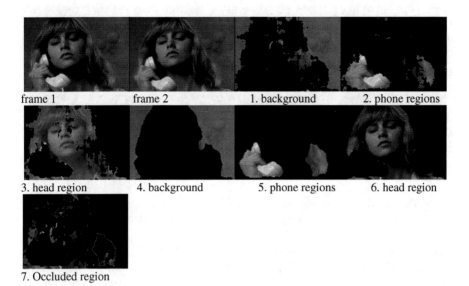

Fig. 5 The segmentation of the Susie calling. 1)-3) are the results of the subspace segmentation method; 4)-6) are the results of the post-processing procedure; 7) the layer boundary goes through the occluded regions.

Moreover, it has been judged that the regions of the phone and the hair patch are in front of the head region. This seems to be a bit strange. In general, the hair patch belongs to the head region. All the hair should be regarded as a whole body on the head and there is no occlusion to each other (unless the hairlines are considered). But it can be observed that the occluded regions overlying on the layer boundary appear at the bottom right of the image, i.e. around the boundaries between the shoulder and the hair. The motion of the hair is independent of that of the shoulder. The shoulder region is classified into the head region. Thus, it is acceptable to preserve this patch as an independent layer as shown in Fig.5(5).

Table Tennis

The table tennis sequence presents a hand holding a table tennis racket to hit a white ball. The image resolution is 190×130 pixels. Applying our motion segmentation method to the two successive frames, we obtained 3 segmented parts of the scene, i.e. ball region, arm region and background region, as shown in Fig.6. The segmentation results by the subspace segmentation approach described in section 2 are shown in Fig.6(1-3). After the post-processing procedure, the segmentation results are shown in Fig.6(4-6), and the region of background is behind those of the ball and arm.

It can be observed that the region of the ball is very small compared to that of the background. In terms of the area comparison strategy, it should be merged into the background. But due to the detected layer edge segments by the Polysegment Alg. in step 3 of the post-processing procedure, the ball region can be preserved.

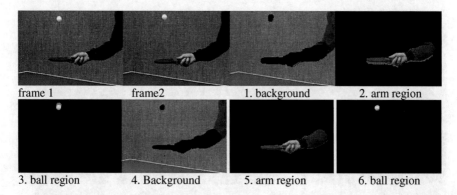

Fig. 6 The segmentation results with 3 groups. 1)-3) are the results by the subspace segmentation method; 4)-6) are the results of the post-processing procedure.

However, this segmentation result is not unique in this sequence. In the other two successive frames, the ball and arm are classified into the same group, which is shown in Fig.7. Although the region of the ball is disconnected to that of the arm, they are still regarded as sharing the same affine motion model. This indicates that motion segmentation cannot guarantee the uniqueness of the solution.

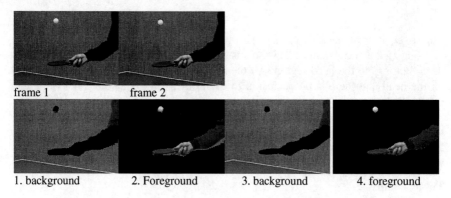

Fig. 7 The segmentation results with 2 groups. 1)-2) are the results of subspace segmentation method; 3)-4) are the results of the post-processing procedure.

5 Conclusion

In this paper, we proposed a novel approach for motion segmentation based on the subspace segmentation techniques. The novelty is that by incorporating the intensity structures of images, our proposed approach can effectively detect the motion layer boundaries and the depth ordering. Different from the previous motion segmentation approaches, our approach provides a non-iterative and global solution to motion segmentation under a unified algebra framework, i.e. the generalized PCA [12,13].

However, it can be noted that our algorithm relies on a given optical flow field. In our experiments, many available optical flow algorithms do not seem suitable for the scenarios with a salient rotation element. This will restrict the applications of our algorithm. It is crucial to further develop a robust optical flow algorithm. Our future work will aim to tackle this challenge.

References

1. Yu, W., Sommer, G., Daniilidis, K.: Multiple motion analysis: in spatial or in spectral domain? Computer Vision and Image Understanding 90(2), 129–152 (2003)
2. Fleet, D.J., Black, M.J., Yacoob, Y., Jepson, A.D.: Design and use of linear models for image motion analysis. Int. J. Comput. Vis. 36(3), 171–193 (2000)
3. Wang, J.Y.A., Adelson, E.H.: Layered representation for motion analysis. In: IEEE Conf. on Computer Vision and Pattern Recognition, pp. 361–366 (June 1993)
4. Szeliski, R., Avidan, S., Anandan, P.: Layer extraction from multiple images containing reflections and transparency. In: IEEE Conf. on Computer Vision and Pattern Recognition, vol. I, pp. 246–253 (June 2000)
5. Black, M.J., Fleet, D.J.: Probabilistic detection and tracking of motion boundaries. Int. J. Comput. Vis. 38(3), 231–245 (2000)
6. Smith, P., Drummond, T., Cipolla, R.: Layered Motion Segmentation and Depth Ordering by Tracking Edges. IEEE Trans. on Pattern Analysis and Machine Intelligence 26(4), 479–494 (2004)
7. Ogale, A.S., Fermuller, C., Aloimonos, Y.: Motion Segmentation Using Occlusions. IEEE Trans. on Pattern Analysis and Machine Intelligence 27(6), 988–992 (2005)
8. Kumar, M.P., Torr, P.H.S., Zisserma, A.: Learning Layered Motion Segmentation of Video. In: Proc. of the Tenth IEEE Int. Conf. on Computer Vision, vol. 1, pp. 33–40 (2005)
9. Irani, M., Anandan, P., Bergen, J., Kumar, R., Hsu, S.: Efficient Representations of Video Sequences and Their Representations. Signal Processing: Image Comm. 8(4), 327–351 (1996)
10. Irani, M., Rousso, B., Peleg, S.: Computing occluding and transparent motions. Int. J. of Computer Vision 2(1), 5–16 (1994)
11. Csurka, G., Bouthemy, P.: Direct Identification of Moving Objects and Background from 2D Motion Models. In: Proc. of IEEE Int. Conf. of Computer Vision, pp. 566–571 (1999)
12. Vidal, R., Ma, Y., Sastry, S.: Generalized Principal Component Analysis (GPCA). IEEE Trans. on Pattern Analysis and Machine Intelligence 27(12), 1–15 (2005)
13. Vidal, R.: Generalized Principal Component Analysis (GPCA): an Algebraic Geometric Approach to Subspace Clustering and Motion Segmentation, Ph.D. Thesis, Electrical Engineering and Computer Sciences, Univer-sity of California at Berkeley (August 2003)
14. Vidal, R., Ma, Y.: A Unified Algebraic Approach to 2-D and 3-D Motion Segmentation. J. of Mathematical Imaging and Vision 25(3), 403–421 (2006)
15. Ogale, A.S., Aloimonos, Y.: A Roadmap to the Integration of Early Visual Modules. Int. J. of Computer Vision 72(1), 9–25 (2007)
16. GPCA matlab codes, http://perception.csl.uiuc.edu/gpca/

17. Papadimitriou, T., Diamantaras, K.I., et al.: Video Scene Segmentation Using Spatial Contours and 3D Robust Motion Estimation. IEEE Trans. On Circuits and Systems For Video Technology 14(4) (2004)
18. Video sequences, http://www.cipr.rpi.edu/resource/sequences/

Appendix

Segmenting Hyperplanes of Dimension K–1 in R^K

Given a set of points $X = \{x^{(j)} \in R^K\}_{j=1}^{N}$ in a homogeneous coordinate system, and linear hyperplanes $\{S_i \subset R^K\}_{i=1}^{n}$ of dimension $k_i = \dim(S_i) = K - 1$, we need to identify S_i. Usually the subspace is given as, $b^{(i)} \cdot x = 0, b^{(i)} \in R^K$. Then, this hyperplane can be represented as,
$S_i = \{x \in R^K : x^T b^{(i)} = 0\}$.

Furthermore, an arbitrary point x lies on one of the hyperplanes if and only if,

$$(x \in S_1) \bigcup ... \bigcup (x \in S_n) \equiv \bigcup_{i=1}^{n} x^T b^{(i)} = 0$$

$$\equiv p_n(x) = \prod_{i=1}^{n} x^T b^{(i)} = 0,$$

$$= y^T c_n = 0$$

where, $y = \left(x_1^n x_2^0 ... x_K^0, x_1^{n-1} x_2^1 ... x_K^0, ..., x_1^0 ... x_{K-1}^0 x_K^n\right)^T \in R^m$ and $c_n \in R^m$ is a coefficient vector consisting of a set of monomials of $\{b^{(i)}\}$ and $m(n, K) = \dfrac{(n+K-1)!}{(K-1)!n!}$.

For a given point set X, we have a linear system on c_n as follows,

$$L_n c_n = \begin{pmatrix} y(x^{(1)})^T \\ ... \\ y(x^{(N)})^T \end{pmatrix} c_n = 0 \in R^N, \tag{A1}$$

where, $L_n \in R^{N \times m}$. When the number of hyperplanes n is known, c_n can be obtained from the null space of L_n. In practice, n is always determined in terms of L_n. For a unique solution of the coefficient vector c_n, it is expected that $rank(L_n) = m(n, K) - 1$, which is a function of variable n. In the presence of noise, let $rank(L_i) = r$ when $n = i$ and $\lambda_{r+1} / \sum_{j=1}^{r} \lambda_j < \varepsilon$, where L_i is the data matrix with $rank = r$, λ_j is the jth singular value of L_i and ε is a given threshold.

Apply GPCA to Motion Segmentation

For any x, we have $p_n(x) = c_n^T y(x)$. Each normal vector $b^{(i)}$ can be obtained from the derivatives of p_n. Consider the derivative of $p_n(x)$ as follows,

$$\nabla p_n(x) = \frac{\partial p_n(x)}{\partial x} = \frac{\partial}{\partial x} \prod_i^n x^T b^{(i)} = \sum_i^n b^{(i)} \prod_{j \neq i} x^T b^{(j)}.$$

For a point $x^{(l)} \in S_l$, $\prod_{j \neq i} b^{(j)^T} x^{(l)} = 0$ for $l \neq i$. It can be noted that there is only one non-zero term in $\nabla p_n(x)$, i.e. $\nabla p_n(x^{(l)}) = b^{(i)} \prod_{j \neq i} b^{(j)^T} x^{(l)} \neq 0$ for $l = i$. Then, the normal vector of S_i is yielded as,

$$b^{(i)} = \frac{\nabla p_n(x^{(l)})}{\|\nabla p_n(x^{(l)})\|}. \tag{A2}$$

In order to get a set of points lying on each hyperplane respectively, so as to determine the corresponding normal vectors $b^{(i)}$, we can choose a point in the given X close to one of the hyperplanes as follows,

$$i = \arg \min_{\nabla p_n(x) \neq 0} \frac{|p_n(x)|}{\|\nabla p_n(x)\|}, \text{ where } x \in X. \tag{A3}$$

After given the normal vectors $\{b^{(i)}\}$, we can classify the whole point set X into n hyperplanes in R^K as follows,

$$label = \arg \min_j \left(x^T b^{(j)} \right), j = 1...n. \tag{A4}$$

This algorithm is called GPCA-PDA Alg. in [12,13].

Polynomial Segmentation Algorithm
Consider a special case of piecewise constant data. Given N data points $x \in R$, we hope to segment them into an unknown number of groups n. This implies that there exist n unknown cluster centers $\mu_1 \neq ... \neq \mu_n$, so that,

$$(x = \mu_1) \cup ... \cup (x = \mu_n),$$

which can be described in a polynomial form as follows,

$$p_n(x) = \prod_{i=1}^n (x - \mu_i) = \sum_{k=0}^n c_k x^k = 0. \tag{A5}$$

To N data points, we have,

$$L_n c \equiv \begin{pmatrix} 1 & x_1 & \cdots & x_1^n \\ \vdots & & & \vdots \\ 1 & x_N & \cdots & x_N^n \end{pmatrix} \begin{pmatrix} c_0 \\ \vdots \\ c_{n-1} \\ 1 \end{pmatrix} = 0, \tag{A6}$$

where $L_n \in R^{N \times (n+1)}$ and $c \in R^{n+1}$. Usually, the group number is estimated as,

$$n = \min\{i : \lambda_{i+1}/\lambda_i < \varepsilon\}, \tag{A7}$$

where λ_i is the ith singular value of L_i, which is the collection of the first $i+1$ columns of L_n, and ε is a given threshold that depends on the noise level.

After solving the coefficient vector c of Eq.(A6), we can compute the n roots of $p_n(x)$, which correspond to the n cluster centers $\{\mu_i\}_{i=1}^n$. Finally, the segmentation of the date is obtained by,

$$i = \arg \min_{j=1,\ldots,n} (x - \mu_j)^2. \tag{A8}$$

The scheme of Eq.(A5-A8) is called as the Polysegment algorithm in [13].

Gait Analysis and Human Motion Tracking

Huiyu Zhou

Abstract. We present a strategy based on human gait to achieve efficient tracking, recovery of ego-motion and 3-D reconstruction from an image sequence acquired by a single camera attached to a pedestrian. In the first phase, the parameters of the human gait are established by a classical frame-by-frame analysis, using an generalised least squares (GLS) technique. The gait model is non-linear, represented by a truncated Fourier series. In the second phase, this gait model is employed within a "predict-correct" framework using a maximum a posteriori, expectation maximization (MAP-EM) strategy to obtain robust estimates of the ego-motion and scene structure, while continuously refining the gait model. Experiments on synthetic and real image sequences show that the use of the gait model results in more efficient tracking. This is demonstrated by improved matching and retention of features, and a reduction in execution time, when processing video sequences.

1 Introduction

One of the applications of closed-circuit television (CCTV) cameras is in surveillance systems that provide monitoring information in order to protect civilians against theft, vandalism and crime [31, 32, 37, 12]. Most of these cameras, to our knowledge, are statically installed in commercial and residential sectors such as schools, banks and parking places. Unfortunately, these cameras can only cover a part of the entire scene, and cannot track subjects that exit the field of view of the cameras. In contrast, portable camera systems can be used to monitor a scene that changes in the field of view of the officer who carries the camera. To conduct effective video analysis, one of the main challenges is efficient recovery of 3-D structure of the scenes and the motion trajectory of the camera, based on 2-D image inputs.

Estimating 3-D structure and camera motion from 2-D images is one of the fundamental problems in computer vision, e.g., [29, 26]. This problem has been

---

Huiyu Zhou
Queen's University Belfast, Belfast, United Kingdom
e-mail: H.Zhou@ecit.qub.ac.uk

continuously investigated and found large amount of real applications. In general, one can adopt two main strategies for the recovery of 3-D structure from passive image data, structure from motion (SFM) and stereo vision [13], each of which relies on the acquisition of video data of the same scene data from different viewpoints. In the case of SFM, one can also recover the ego-motion of the sensor with respect to a world coordinate system, performing 3-D scene reconstruction and navigation [38, 35]. This implies knowledge of corresponding locations in the several images. Lepetit and Fua [17] have described the general principles of feature detection, tracking and 3-D reconstruction, and Oliensis [22] gave an earlier detailed critique of the comparative strengths and weaknesses of several, well-documented approaches to SFM. The majority of existing methods for feature tracking use frame-to-frame prediction models, based for example on Kalman filter [14, 20], particle filtering [25], and optimisation based approaches [33, 21, 34]. One of the numerous examples is the MonoSLAM system developed by Davison et al. [8], who utilised a probabilistic feature-based map that represents a snapshot of the current estimates of the state of the camera and the overall feature points. This feature map was initialised at system start-up and updated by the extended Kalman filter that considered frame-by-frame computation. The state estimates of the camera and feature points were also updated during camera motion.

In spite of their success in certain applications, these established approaches do not take into account the long-term history of the camera motion. In contrast, we track features and recover ego-motion and 3-D structure from a temporal image sequence acquired by a single camera mounted on a moving pedestrian. Our contribution is to show that the use of an explicit longer term, non-linear human gait model is more efficient in this case. Fewer features are lost and the processing time per frame is lessened as either the search window or the frame rate can be reduced.

Our work was motivated by the study reported in [19], where Molton *et al.* used a robotic system to make measurements on the gait of a walking person, while a digital compass and an inclinometer were used to record rotation. An iterated and extended Kalman filter was used in their work to initialise the wavelet parameter estimates, then running across the whole period of activity. In our work, the motion is computed directly from the video data, and as already stated, the emphasis is on the use of a longer term model to increase algorithmic efficiency. In what follows, we use the term "ego-motion" to refer to both frame-to-frame and longer-term periodic motion. The expression "camera transformation" refers to the ego-motion of the camera between any two frames, and the expression "gait model" refers to the longer-term ego-motion of the camera over many frames.

In Section 2 we give an overview of our approach, which has two phases, initialisation and continuous tracking. We then expand on the key components of our strategy in Sections 3 and 4. In Section 3 we discuss the GLS method to recover the scene geometry and ego-motion within Phase 1, effectively finding structure from motion (SFM) [36]. In Section 4 we discuss the formulation of the dynamic gait model within the MAP-EM framework of Phase 2 to continually perform SFM with improved efficiency while periodically updating the gait parameters. In Section 5, we present results that show the improved efficiency in comparison with standard methods. Finally, in Section 6 we summarize our findings.

2 An Overview of the Proposed Approach

The proposed approach has two phases, *initialisation* and *continuous tracking*. Assume that m frames are included in the initialisation phase and n frames ($n > m$) in the whole sequence. The intrinsic camera parameters are known from previous calibration. The purpose of the first phase is to acquire the long-term gait model of the pedestrian.

Phase 1: Initialisation

Of the well-tested feature tracking algorithms, reviewed by Lepetit and Fua [17], the Shi-Tomase-Kanade (STK) tracker [27] is used because it is still one of the most accurate and reliable. The STK tracker has two stages, feature selection and feature tracking. The feature selection process computes the eigenvalues of a gradient function for each pixel, comparing the result with a fixed threshold. In the published algorithm, image features with higher eigenvalues were considered as good features to track. However, we use the SUSAN (Smallest Univalue Segment Assimilating Nucleus) operator instead to select an initial feature set in the first frame, due to its known immunity to noise [28]. This is not used subsequently, unless the number of tracked features falls below a pre-defined threshold in which case we are able to re-initialise with new features.

```
Select C corner features in frame 1
For frames i=1:1:m-1 (STK-GLS)
  Match features across frames i and
  (i+1);
  Estimate fundamental matrix;
  Refine list of matched features;
  Recover camera transformation and
  scene geometry;
EndFor
Fit periodic gait model to camera trans-
formation data from frames 1 to m
```

As we track these features in successive frames, we minimise the residue error using an affine transformation, as in the STK algorithm, assuming the displacement is small. However, we use a generalised least squares (GLS) algorithm [42] to recover the frame-to-frame camera transformation and scene geometry as this is a robust estimation technique that can derive good results when the error distribution is not normally distributed [40]. Typically, m is chosen large enough to recover about two complete strides, say $m = 50$ frames for a 25Hz sampling rate. The number of features, C, is a user parameter, typically set to 150. This leads to the recovery of a temporary motion model with six degrees of freedom, i.e. the 3 displacements and

3 Euler angles, which can be stored in a gait library using a truncated Fourier series representation.

Phase 2: Continuous tracking (IRLS: iteratively re-weighted least squares)

```
For frame i=m:m0:n (Proposed MAP-EM)
  Predict the feature positions in the
  (i+m0)th frame using the gait model;
  Apply a coarse-to-fine strategy to
  match the features;
  Compute frame-to-frame transformation
  and scene geometry on every m'th frame;
    Update the periodic gait model using
    IRLS;
EndFor
```

In Phase 2, we use and update the gait model to improve the prediction of the location of features in each new frame. Since this prediction is based on a longer history, not every frame needs to be considered so we include a parameter $m0$ as the gap between frames. (In the experimental section $m0$ is varied so that the performance of the proposed algorithm can be fully evaluated). The coarse-to-fine matching parameters, e.g., window size, are selected according to the variance of the Euclidean distance between the predicted and measured feature positions. A maximum a-priori, expectation-maximization (MAP-EM) algorithm is used to find the best feature match and determine the camera transformation parameters between two frames. The gait model is examined and updated if necessary on a periodic basis using iteratively re-weighted least squares (IRLS) [2]. To maintain its validity, the m most recent frames are used, applying *Spearman correlation* to compare the recent gait parameters individually with the stored values. If the correlation coefficients are lower than a pre-defined threshold (typically 0.9), then the current gait model is updated. Accumulation of errors in the motion tracking can deteriorate the performance of the proposed tracker. In order to reduce this accumulated error we can re-estimate the overall parameters by collecting very similar (or the same) images that were just used. For example, multiple estimates of the position of a known 3-D point should be identical. This strategy works effectively if the iteration runs several times.

3 Phase 1: Establishing the Gait Model

In Phase 1, we recover the fundamental matrix, F, between each pair of successive frames in the sequence. F defines a transformation from a point $\mathbf{p}_1$ in one image to the epipolar line $\mathbf{l}_2$ in the other image,

$$\mathbf{l}_2 = F\mathbf{p}_1, \tag{1}$$

or between corresponding points in the two images,

$$\mathbf{p}_2^T F \mathbf{p}_1 = 0, \qquad (2)$$

where the positions of points $\mathbf{p}_1$ and $\mathbf{p}_2$ are defined in image coordinates. We solve the equation, $\mathbf{x}^T \mathbf{a} = 0$, where $\mathbf{a}$ is a 9-vector including the entries of the fundamental matrix F, and $\mathbf{x}^T$ is a 1×9 parameter matrix whose elements come from those elements of $\mathbf{p}_1$ and $\mathbf{p}_2$. Solutions for F are well understood [36], but in our case outliers may be caused by incorrect image correspondences. In general, errors embedded in the sum of the squared residuals of Euclidean distance from image points to their corresponding epipolar lines are *non-Gaussian*, as illustrated in the images and histogram of Fig. 1. Therefore we employ a two-stage, *generalised least squares* model [42]. This is an extended version of linear mixed-effects models that have been used for the analysis of balanced or unbalanced grouped data, such as longitudinal data, repeated measures, and multilevel data [24, 16].

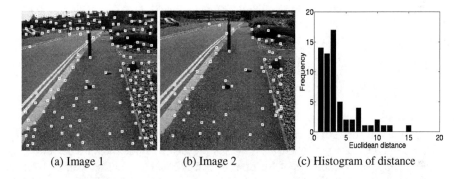

(a) Image 1   (b) Image 2   (c) Histogram of distance

Fig. 1 An example demonstrating the distribution of the squared residuals of Euclidean distance: (a) and (b) a pair of real images with the corner features superimposed, and (c) the frequency (point numbers) versus Euclidean distances (pixels), where the width of each bin is 1 pixel.

For simplicity, the general form of the Gauss-Markov linear model is considered. Let $\mathbf{Y}_i$ be a 9-vector with index i (each element of $\mathbf{Y}_i$ is -1):

$$\mathbf{Y}_i = \mathbf{X}_i \mathbf{A}_i + \varepsilon_i, \quad i = 1, 2, ..., N, \qquad (3)$$

where $\mathbf{X}_i$ is a 9×8 parameter matrix (9 random samples from the corresponding points and 8 unknown elements of the F matrix), $\mathbf{A}_i$ is an 8-vector including the unknown elements of the F matrix (the ninth is 1 given a normalised F matrix), and ε_i is an independent error. The dimension of $\mathbf{X}_i$ is designed such that the computation of the F matrix becomes more efficient and robust [40]. For variable index i,

$\mathbf{A}_i$ indicates individual fundamental matrices F_i determined by the ith subsampling group. Second, the coefficients $\mathbf{A}_i$ are associated with the population covariates by

$$\mathbf{A}_i = \mathbf{v}_i \mathbf{b} + \gamma_i, \qquad (4)$$

where $\mathbf{v}_i$ is an 8×8 population design matrix, $\mathbf{b}$ is an 8-vector, γ_i is a vector of random effects independent of ε_i. An optimal solution for Eq. (3) leads to the minimisation of ε_i.

To obtain a solution for $\mathbf{A}_i$, the two-stage strategy is combined with a moment estimator, based on the optimisation technique proposed by Demidenko and Stukel [9]. The first stage refers to Eq. (3), while the second stage contributes to Eq. (4). The moment estimator is used to compute the variance between the measurements and the estimates against the number of the remaining samples. Omitting intermediate steps that can be found in their paper, we have:

$$\mathbf{A}_i = (\sum_i \mathbf{Z}_i^T \mathbf{H}_i^{-1} \mathbf{Z}_i)^{-1} \sum_i \mathbf{Z}_i^T \mathbf{H}_i^{-1} \mathbf{C} \tilde{\mathbf{A}}_i, \qquad (5)$$

where $\mathbf{A}_i$ is an unbiased solution to the fundamental matrix. $\mathbf{C} = \mathbf{Z} = \mathbf{I}$ (identity matrix), and $\tilde{\mathbf{A}}_i$ and $\mathbf{H}$ can be computed using the numerical technique introduced in [42]. The proposed linear mixed-effects model is not robust to the existence of outliers that perturb the optimization by introducing ambiguities. To make the proposed strategy robust, we investigate the squared residuals of the Euclidean distances from image points to their corresponding epipolar lines in accordance with the two-stage estimator [39].

A confidence interval, μ, is used to determine the probability of inliers (valid correspondences) or outliers (rogue points). This confidence interval is constructed as $a_m - t_m s \leq \mu \leq a_m + t_m s$, where a_m is the sample mean, s is the standard deviation, and t_m depends on the degrees of freedom, which is equal to one less than the size of a random subsample, i.e. eight, and the level of confidence. To determine t_m, the t distribution is used. The t distribution is an infinite mixture of Gaussians, and is often employed as a robust measure to "correct" the distorted estimation created by a few extreme measurements.

$$\mu \approx a_m \pm 2.896 \times \frac{1}{M-1} \sqrt{\sum_i (\tilde{d}_i - a_m)^2}, \qquad (6)$$

Assuming there are i correspondences across images, the confidence interval of the distribution is determined by Eq. (6), where $a_m = \frac{\tilde{d}_i}{M}$, M is the number of the image correspondences, and $\tilde{d}_i$ are the residuals computed by applying the fundamental matrix estimated in the last subsection to the overall image correspondences, and 2.896 is based on a significance level of 0.01 and the 8 degrees of freedom (with reference to the upper critical values of the student's t distribution). Examples of the epipolar geometry estimation can be found in Fig. 2. Once the fundamental matrix has been obtained, the essential matrix, E, can be obtained using the known

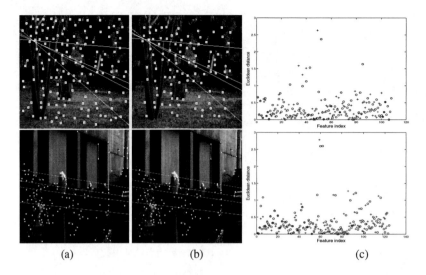

Fig. 2 Estimation of epipolar geometry: (a) and (b) matching results and epipolar geometry recovered, and (c) the Euclidean distances (pixels) versus feature index.

intrinsic camera matrix M_{int}. The camera transformation between the two images and the scene geometry is recovered by assuming a pinhole camera model [36].

It is known that walking is an internally periodic process that becomes mature at around 12 months [30]. Let $d(t)$ be the periodic component of the generic displacement function in terms of time t during motion. The "oscillating property" of the motion pattern is expressed as a truncated Fourier series

$$d(t) = d_0 + \sum_{k=1}^{N} d_k \sin(\omega_k t + \phi_k), \qquad (7)$$

where the frequencies are $\omega_k = \frac{2\pi}{T_p^k}$ (T_p^k is a time period). ω_k indicate the k-th harmonics of the function $d(t)$, i.e. the first k components of the Fourier series after the fundamental frequency. d_0 is the mean value, and d_k and ϕ_k are the amplitude and phase of the k-th harmonic in one stride period respectively [4]. As demonstrated in the experimental results presented here, and more extensively in [40], a value of $N = 3$ allows normal gait patterns to be properly represented. Examples can be found in Fig. 3. A model of lower complexity leads to large fitting errors and of higher complexity to fitting of random and aperiodic fluctuations. Hence, for six degrees of freedom, each represented by ten parameters in a truncated series, the motion model has a total of 60 parameters. The camera transformations obtained from successive frame-to-frame correspondences are used to form the gait model, established over the sequence. As discussed in Section 2, a change in ω_k requires updating of the gait model. This deterministic gait model is used to achieve better prediction of the movement of features within the camera field of view over a

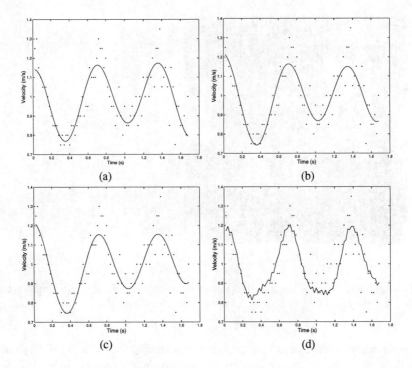

Fig. 3 Fitting a group of speed values using different methods: (a) the Nelder-Mead method; (b) the Levenberg-Marquardt method (3-order); (c) the M-estimator; and (d) the Gauss-Newton method. Dots: real velocity data; solid lines: fitted curves.

longer period of time and so improve the efficiency of gait based ego-motion tracking, which is the principal contribution of this paper. Our use of a periodic motion model to represent gait is similar to Molton *et al.* [19], but we derive the gait model directly from the images, thus relieving the pedestrian of additional instrumentation (compass and inclinometer). We have also extended the representation to a truncated series of sinusoidal terms for greater accuracy. As defined, the methodology is not general but gait-specific because of the form of Eq. (7), the principle of using a longer term motion model could be applied to other examples of periodic motion. Here, the gait model is constrained by the fact that the translation and rotation of the camera between two neighboring frames are less than 0.1 m and 10 degrees, respectively.

4 Phase 2: Recovering Ego-Motion and Scene Structure Using a Dynamic Gait Model

The continuous tracking phase has two main stages: prediction and correction. If the prediction yielded by the gait model is accurate, then this leads either to a smaller search window, or allows a longer gap between frames, each resulting in greater efficiency. Further, there is less chance of an incorrect match as the probability of similarity between detected features is reduced. The nature of the gait model described by Eq. (7) determines the use of a nonlinear optimization process. This is a *dependent* multivariate process, and the stability of the possible solution is not guaranteed since too many unknown parameters reside in the optimization [15].

We register the projection of a 3-D "template", yielded by the gait model, with the 2-D feature points in the latest frame. This is an incomplete data problem because the correspondences are not known accurately *a priori* as there are incorrect correspondences and errors in position. We expect that a number of *outliers* may occur in the corresponding matches. Hence, we derive a *global* rather than a *local* similarity between the actual correspondences and the image projections of the scene structure. In the presence of incorrect feature correspondence, robust registration and recovery of the camera transformation between frames is explored using a maximum a posteriori (MAP) strategy, instantiated by the expectation-maximization (EM) algorithm [11]. Hence, we iterate the *expectation and maximization* until convergence is reached at the *global minimum*. The expectation step indicates *a posteriori* probabilities of the incomplete data using Gaussian mixture models, given the image observations and belief in motion provided by the gait model. The maximization step involves a *maximum a posteriori estimate* to refine the predicted motion parameters in order to obtain a minimum sum of Euclidean distances between image points. Our approach is inspired by that used by Cross and Hancock [7], Choi et al. [5] and Zhou et al. [41]. However, the key difference is the use of the gait model to predict frame-to-frame camera transformations, thus improving the efficiency of the approach. In comparison with [41], we have extended our evaluation to include synthetic and real pedestrian sequences in which walking velocity changes. Furthermore, we have added an experiment investigating the effect of moving obstacles (other pedestrians) in real image sequences.

4.1 Estimating Motion Parameters by a MAP Strategy

Consider a *dynamic* representation for the registration, $f(\tilde{\alpha}_t, \tilde{\beta}_t, \tilde{\phi}_t)$, where $\tilde{\alpha}_t$ refers to the 3-D points recovered from corresponding image points, $\tilde{\beta}_t$ is the image observation, and $\tilde{\phi}_t$ is the current prediction of the camera transformation, based on the longer term gait model, at time t. Given a good $\tilde{\phi}_t$, the posterior probability (or the likelihood of the hypothesis, $\tilde{\phi}_t$), $p(\tilde{\beta}_t|\tilde{\phi}_t)$, is maximized to find an optimal $\tilde{\beta}_t$.

Assuming individual image points are conditionally *independent* [5], the joint probability is therefore

$$p(\tilde{\beta}_t|\tilde{\phi}_t) = \prod_i p(\tilde{\beta}_{ti}|\tilde{\phi}_t), \qquad (8)$$

where i is the index of an image point. Using Bayes' rule,

$$p(\tilde{\beta}_{ti}|\tilde{\phi}_t) = \sum_j p(\tilde{\beta}_{ti}|\tilde{\alpha}_{tj},\tilde{\phi}_t) p(\tilde{\alpha}_{tj}), \qquad (9)$$

where j is also an image point index. The maximum log-likelihood estimate of $\tilde{\phi}_t$ is explicitly defined as follows

$$L(\tilde{\phi}_t) \equiv \log p(\tilde{\beta}_t | \tilde{\phi}_t). \tag{10}$$

The correspondences between features are not known a-priori from frame to frame in spite of the capture of a *concrete* gait model to generate $\tilde{\phi}_t$, which is used to generate a 3-D model $\tilde{\alpha}_t$. However, the conditional logarithm of the posterior probability, $Q(\tilde{\phi}_t | \tilde{\phi}_{t-1})$, can instead be computed over the previous estimates of $\tilde{\phi}_t$. Substituting Eq. (9) into Eq. (10) and considering the Markovian properties, one can re-write the conditional log-likelihood as

$$Q(\tilde{\phi}_t | \tilde{\phi}_{t-1}) = \sum_i \sum_j p(\tilde{\alpha}_{t j} | \tilde{\beta}_{t i}, \tilde{\phi}_{t-1}) \log p(\tilde{\beta}_{t i} | \tilde{\alpha}_{t j}, \tilde{\phi}_t). \tag{11}$$

4.2 EM Algorithm

As applied here, the EM algorithm starts from an initial "guess" of the scene structure, which is derived from the motion parameters provided by the gait model and previous feature positions, and then projects the perspective 3-D "template" to a 2-D image using the features to be matched across two frames of the sequence using the following iterative steps:

1. **E-step**:

In this step, we formulate the *a posteriori* probability of the incomplete data set, $p(\tilde{\alpha}_{t j} | \tilde{\beta}_{t i}, \tilde{\phi}_{t-1})$, contained in Eq. (11). Bayes' rule is again applied to obtain

$$p(\tilde{\alpha}_{t j} | \tilde{\beta}_{t i}, \tilde{\phi}_{t-1}) = \frac{p(\tilde{\alpha}_{t j} | \tilde{\phi}_{t-1}) p(\tilde{\beta}_{t i} | \tilde{\alpha}_{t j}, \tilde{\phi}_{t-1})}{\sum_k [p(\tilde{\alpha}_{t k} | \tilde{\phi}_{t-1}) p(\tilde{\beta}_{t i} | \tilde{\alpha}_{t k}, \tilde{\phi}_{t-1})]}. \tag{12}$$

To pursue solutions for $p(\tilde{\alpha}_{t j} | \tilde{\phi}_{t-1})$ and $p(\tilde{\beta}_t | \tilde{\alpha}_{t j}, \tilde{\phi}_{t-1})$ in Eq. (12), the probability $p(\tilde{\alpha}_{t j} | \tilde{\phi}_{t-1})$ may be written as

$$p(\tilde{\alpha}_{t j} | \tilde{\phi}_{t-1}) = \frac{1}{N} \sum_i p(\tilde{\alpha}_{t j} | \tilde{\beta}_{t i}, \tilde{\phi}_{t-1}), \tag{13}$$

where N is the number of the features. This indicates that the posterior probability $p(\tilde{\alpha}_{t j} | \tilde{\phi}_{t-1})$ is determined by the individual joint densities $p(\tilde{\alpha}_{t j} | \tilde{\beta}_{t i}, \tilde{\phi}_{t-1})$ over the feature points considered. To compute $p(\tilde{\alpha}_{t j} | \tilde{\phi}_{t-1})$, it is necessary to take into account the initial estimate of $p(\tilde{\alpha}_{t j} | \tilde{\beta}_{t i}, \tilde{\phi}_{t-1})$, which seriously affects the characteristics of convergence, e.g. accuracy and efficiency, for final correspondence. Here, we preset it to be $\frac{1}{N}$. In other words, the posterior probability of registration of each feature is uniform in the first instance.

Gait Analysis and Human Motion Tracking

As stated, we assume a multivariate Gaussian motion model for the conditional probability $p(\tilde{\beta}_{ti}|\tilde{\alpha}_{tj},\tilde{\phi}_{t-1})$. So, its maximum likelihood estimate can be represented by the continuous density [10] as

$$p(\tilde{\beta}_{ti}|\tilde{\alpha}_{tj},\tilde{\phi}_{t-1}) = \left(\frac{1}{2\pi}\right)^{\frac{D_z}{2}} \left(\frac{1}{det(\Sigma)}\right)^{\frac{1}{2}} exp\left(-\frac{1}{2}\varepsilon\Sigma^{-1}\varepsilon^{\mathbf{T}}\right), \quad (14)$$

where T indicates transpose; $D_z = 6$ since there are six transformation parameters; the error-residual

$$\varepsilon = \tilde{\beta}_{ti} - \gamma_{tj}, \quad (15)$$

which is the Euclidean distance between the matched image feature, $\tilde{\beta}_{ti}$, and the projected position derived from the gait model and prior structure, γ_{tj}. The variance-covariance matrix Σ and its inverse Σ^{-1} are both $2N \times 2N$ positive definite symmetric because of their elements σ_{ij} from the covariance of ε_i and ε_j, i.e.

$$\Sigma = E[\varepsilon\varepsilon^{\mathbf{T}}]. \quad (16)$$

Re-writing the logarithmic part of Equation (11) results in the new form as follows:

$$Q(\tilde{\phi}_t|\tilde{\phi}_{t-1}) = -\frac{1}{2}\sum_i\sum_j p(\tilde{\alpha}_{tj}|\tilde{\beta}_{ti},\tilde{\phi}_{t-1})(\vartheta - (\tilde{\beta}_{ti} - \gamma_{tj})(\tilde{\beta}_{ti} - \gamma_{tj})^T). \quad (17)$$

where ϑ stands for the total summation of other terms after taking logarithms. The 3-D position of γ_t is

$$\tilde{\alpha}_t = \mathbf{R}(t)\tilde{\alpha}_{t-1} + \mathbf{T}(t), \quad (18)$$

where $\mathbf{R}(t)$ (or $\mathbf{R}$ afterwards) is a rotation matrix represented by Euler angles, and $\mathbf{T}_d(t)$ (or $\mathbf{T}_d$ afterwards) is a translation vector which contains T_x, T_y, and T_z. The updated parameters for the tri-axial Euler angles, and displacements can be expressed as

$$\begin{cases} \theta_x(t) = \theta_{x0} + \sum_{k=1}^{3} \theta_{x_k} \sin(\omega_{r_xk}t + \phi_{r_xk}), \\ \theta_y(t) = \theta_{y0} + \sum_{k=1}^{3} \theta_{y_k} \sin(\omega_{r_yk}t + \phi_{r_yk}), \\ \theta_z(t) = \theta_{z0} + \sum_{k=1}^{3} \theta_{z_k} \sin(\omega_{r_zk}t + \phi_{r_zk}), \\ T_x(t) = T_{x0} + \sum_{k=1}^{3} T_{x_k} \sin(\omega_{t_xk}t + \phi_{t_xk}), \\ T_y(t) = T_{y0} + \sum_{k=1}^{3} T_{y_k} \sin(\omega_{t_yk}t + \phi_{t_yk}), \\ T_z(t) = T_{z0} + \sum_{k=1}^{3} T_{z_k} \sin(\omega_{t_zk}t + \phi_{t_zk}), \end{cases} \quad (19)$$

where the overall coefficients, i.e. $\theta_{x0}, \omega_{r_xk}$, etc., can be estimated using the iteratively-reweighted least-squares (*IRLS*) method [2] with historic data. *IRLS* continuously updates a weight function so as to minimise the effects of gross outliers within the optimization. The motion parameters at time instant, t, will become deterministic.

2. **M-step**:

Eq. (17) is iterated in order to find its maximum. This involves finding $\tilde{\phi}_t = M(\tilde{\phi}_{t-1})$ so that

$$Q(\tilde{\phi}_t|\tilde{\phi}_{t-1}) \geq Q(\tilde{\phi}_{t-1}|\tilde{\phi}_{t-1}). \quad (20)$$

To seek a fast solution to this maximisation problem, we can differentiate $Q(\tilde{\phi}_t|\tilde{\phi}_{t-1})$ in terms of the six transformation parameters, and set them to be zero individually. Therefore, the optimal solutions for the six DOFs can be obtained by solving these numerical equations:

$$\frac{\partial Q(\tilde{\phi}_t|\tilde{\phi}_{t-1})}{\partial \mathbf{T}_d} = 0, \tag{21}$$

which is available, and

$$\frac{\partial Q(\tilde{\phi}_t|\tilde{\phi}_{t-1})}{\partial \theta} = \frac{\partial Q(\tilde{\phi}_t|\tilde{\phi}_{t-1})}{\partial \mathbf{R}} \frac{\partial \mathbf{R}}{\partial q} \frac{\partial q}{\partial \theta} = 0, \tag{22}$$

where the rotational vector θ has three Euler angles, $(\theta_x, \theta_y, \theta_z)$. Further,

$$\frac{\partial Q(\tilde{\phi}_t|\tilde{\phi}_{t-1})}{\partial \mathbf{R}} = -\sum_i \sum_j p(\tilde{\alpha}_{tj}|\tilde{\beta}_{ti}, \tilde{\phi}_{t-1})(\tilde{\beta}_{ti} - \gamma_{tj}) f \tilde{\alpha}_{xt} / \tilde{\alpha}_{zt}, \tag{23}$$

The rotation matrix can be formulated using a unit *quaternion* that can be expressed as $q = (q_0, q_1, q_2, q_3)$:

$$\mathbf{R} = \begin{bmatrix} q_0^2 + q_1^2 - q_2^2 - q_3^2 & 2(q_1q_2 - q_0q_3) & 2(q_1q_3 + q_0q_2) \\ 2(q_1q_2 + q_0q_3) & q_0^2 - q_1^2 + q_2^2 - q_3^2 & 2(q_2q_3 + q_0q_1) \\ 2(q_1q_3 - q_0q_2) & 2(q_2q_3 + q_0q_1) & q_0^2 - q_1^2 - q_2^2 + q_3^2 \end{bmatrix}. \tag{24}$$

The Levenberg-Marquardt (L-M) technique [23] is applied to search for the optimal solutions, as this is a non-linear optimisation problem, solved by combining gradient descent and Gauss-Newton iteration. The variation of the rotation matrix from frame to frame can be derived using a linear optimisation [3], where the incremental rotation quaternion is computed at each frame with

$$\begin{cases} \Delta q = (\sqrt{1-\zeta}, \theta_x/2, \theta_y/2, \theta_z/2), \\ \zeta = (\theta_x^2 + \theta_y^2 + \theta_z^2)/4. \end{cases} \tag{25}$$

Assuming a vector $v = [v_1, v_2, v_3]$, then the derivatives of the rotation matrix with respect to the quaternion can be obtained

$$\frac{\partial \mathbf{R}^{-1}}{\partial q} = \begin{bmatrix} c_1 & c_4 & -c_3 & c_2 \\ c_2 & c_3 & c_4 & -c_1 \\ c_3 & -c_2 & c_1 & c_4 \end{bmatrix}, \tag{26}$$

where

$$\begin{cases} c_1 = q_1 v_1 + q_4 v_2 - q_2 v_3, \\ c_2 = -q_3 v_1 + q_0 v_2 + q_1 v_3, \\ c_3 = q_2 v_1 - q_1 v_2 + q_0 v_3, \\ c_4 = q_1 v_1 + q_2 v_2 + q_3 v_3. \end{cases} \tag{27}$$

(refer to [6]). The vector v satisfies the following equations:

$$\begin{cases} |v| = \theta_{deg}, \\ \cos(0.5\theta_{deg}) = \sqrt{1-\zeta}, \end{cases} \tag{28}$$

where θ_{deg} is an intermediate parameter used in the process.

The derivation of the transformation parameters using the EM scheme is based on the prediction of the gait model. Therefore, the final variation of the motion between frames is the arithmetic summation of the prediction and the changes due to the optimisation. The EM platform should efficiently converge to a unique solution, ϕ^*, rather than several solutions due to under-estimation or over-estimation, where a large inaccuracy arises. It is difficult to prove this in theory. In practice, Meilijson [18] claims that Newton-type or other gradient methods provide the solution required quickly but tend to be unstable. Therefore we apply the Levenberg-Marquardt method [23], using the prediction based on the history of the recovered motion parameters.

5 Experimental Results

To demonstrate improved performance, we compare the proposed gait-based ego-motion tracking with the benchmark STK algorithm, which uses a short-term displacement model. We employ synthetic test data, from a checked test pattern and a computer game simulation for comparison of the algorithms with known periodic gait parameters of the form of Eq. (7). We also employ real data from a camera mounted on a pedestrian. As we do not have independent extraction of gait in this case, we compare texture mapped images using extracted parameters with the real image data, which gives a strong subjective comparison. The experimental results are augmented considerably in [40].

5.1 Synthetic Checked Target: Accuracy Tests

The algorithm was tested using five synthetic sequences of a checked pattern, illustrated in Fig. 4 (frame size: 630 × 630 *pixels²*), including translational and rotational motions of varying velocity. A right-handed Cartesian coordinate system was used in which the Z axis is normal to the image plane. Although the target was simple, we added progressively increasing levels of Gaussian noise of mean zero and variance 0.0 to 12.0 on an intensity scale from 0.0 to 255.0. 150 features were extracted in the first frame but this number was reduced as features were lost as tracking continued. In the sequence of Fig. 4, the pattern translates along the X and Z axes and rotates about the Z-axis. In phase 1, the gait model is established in the first 50 frames using the STK algorithm, Fig. 5 shows that the recovery of the t_x parameter is degraded with increased noise; for a variance of 12.0, the estimated translation along the X axis has an absolute error of 0.1 - 0.5 units (1 unit: 19cm). Similar results are observed for the t_z and θ_z components [40]. The key comparison

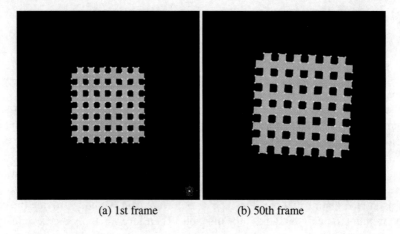

(a) 1st frame (b) 50th frame

Fig. 4 Synthetic images with the feature points superimposed

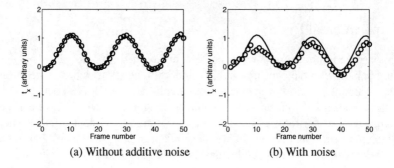

(a) Without additive noise (b) With noise

Fig. 5 Combination of translation and rotation (the arbitrary units are relative to unknown parameters and hereafter). Estimating t_x in the sequence with random noise added using the STK-based approach: (a) no additive noise, (b) (0.0,12.0) noise. "-" - ground truth, and "o" - STK-based estimates.

is to compare results for tracking in phase 2 using the gait model against the benchmark STK algorithm from frame 51 onwards. The estimates of the translational (t_x and t_z) and rotational (θ_z) components of motion with added noise of variance 12.0 are illustrated in Fig. 6. Table 1 provides error statistics for the accuracy of feature location. The mean and standard deviation of the errors in all three estimated parameters are reduced in most cases when using the gait model, as is also apparent from Fig. 6 since the gait data, marked with "×" symbols, is closer to the ground truth marked by the solid line. Hence, the data demonstrates quantitatively that the gait-based technique recovers the motion parameters and locates the features more accurately.

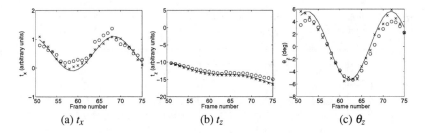

(a) t_x (b) t_z (c) θ_z

Fig. 6 Comparison of the ground-truthed motion and estimation using the gait-based and the STK-based tracking approaches with (0.0,12.0) noise added: (a) t_x, (b) t_z, (c) θ_z. "-" - ground truth, "o" - STK-based estimates, and "×" - gait-based estimates.

Table 1 Ego-motion tracking statistics for different noise levels: where m_s and σ_s are the absolute mean and RMS errors by the STK tracker; m_g and σ_g are the absolute mean and RMS errors by the gait-based scheme. Units: displacements - arbitrary and angles - degrees.

Noise	t_x	t_z	θ_z
Var	m_s σ_s m_g σ_g	m_s σ_s m_g σ_g	m_s σ_s m_g σ_g
0	0.08 0.11 0.02 0.03	0.24 0.35 0.09 0.12	0.27 0.41 0.35 0.52
2	0.06 0.06 0.03 0.03	0.08 0.13 0.04 0.07	0.46 0.65 0.13 0.22
4	0.02 0.02 0.01 0.01	0.04 0.05 0.04 0.04	0.73 0.88 0.48 0.61
6	0.32 0.41 0.16 0.24	0.77 0.91 0.83 0.96	0.64 0.83 0.41 0.57
8	0.01 0.02 0.01 0.01	0.32 0.45 0.21 0.39	0.58 0.72 0.38 0.43
12	0.16 0.21 0.03 0.05	0.55 0.67 0.13 0.18	0.74 0.85 0.51 0.59

5.2 Synthetic Checked Target: Changing Sample Rates

The purpose of these experiments was to see if we could gain greater efficiency of tracking by using the gait model to reduce the sampling rate, thus reducing the computational load. Examining Table 2 and Fig. 7(a), sampling every third frame, errors in the estimates of t_z using the gait model are reduced slightly in noise-free data. However, when noise is added in Fig. 7(b), the gait model is necessary to retain tracking as features are lost during tracking in the basic STK method. For example, previous to frame 21, the estimates of t_z using the STK tracker are displaced above the correct values because the fundamental matrix estimation is sensitive to errors that occur in point correspondence. At frame 71, the step function is caused by newly introduced incorrect correspondences that continue to propagate. Estimates of rotation in Table 2 and Fig. 8 are more complicated. Although the gait-based method again performs better than the STK-based technique, the differences are not as marked as with the t_z and t_x parameters. They are almost comparable at a variance of additive Gaussian noise of 12.0. Nevertheless, the absolute errors using gait are less than 1.2 degrees on average. Taken overall the use of the gait model does improve the accuracy of the recovered parameters and allows a reduced sampling rate for the synthetic data. In practice, the gait model allowed recovery of transformation

Table 2 Ego-motion tracking statistics for different noise levels when sample rates are decreased (notation is same as Table 1).

Noise Variance	t_x m_s σ_s m_g σ_g	t_z m_s σ_s m_g σ_g	θ_z m_s σ_s m_g σ_g
0	4.02 4.11 0.04 0.05	0.06 0.07 0.06 0.06	0.52 0.71 0.26 0.47
2	4.26 4.31 0.03 0.03	0.03 0.03 0.03 0.03	0.71 0.83 0.37 0.47
4	6.74 6.87 0.02 0.03	1.72 1.75 0.08 0.09	0.82 0.94 0.51 0.72
6	4.95 5.03 0.22 0.23	0.16 0.18 0.17 0.17	0.73 0.79 0.61 0.73
8	4.60 4.62 0.28 0.34	0.11 0.13 0.09 0.10	0.58 0.72 0.38 0.43
12	10.78 12.81 0.35 0.41	10.84 15.19 0.33 0.38	0.78 0.94 0.58 0.67

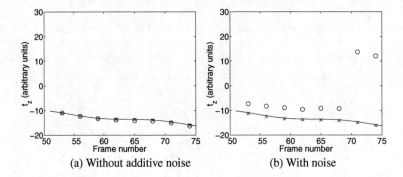

Fig. 7 Combination of translation and rotation (t_z). Comparison of the ground-truthed motion and estimation using the gait-based and the STK-based approaches with random noise added: (a) no additive noise, (b) (0.0,12.0) noise. "-" - ground truth, "o" - STK-based estimates, and "×" - gait-based estimates.

parameters at sampling rates of up to every sixth frame on this sequence, but it was not possible to make a comparison because the basic tracker was unable to recover the motion parameters.

5.3 Synthetic Human Motion Sequences

In the next set of experiments, we extended the evaluation to a full 3-D environmental "walk-through", using a computer animation package to provide synthesis of translational and rolling gait. This environment is far more realistic than the checked target, but we can still compare the recovered parameters against ground truth as the gait parameters can be fully programmed using the "Quake" engine, which is publicly available at http://www.codeproject.com/managedcpp/quake2.asp. We used a demonstration video ("indoors navigation") in which the "observer" wanders in a castle in changing illumination, which together with the more complex 3-D environment can cause more missed feature correspondences than with the checked target.

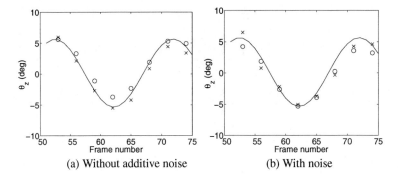

Fig. 8 Combination of translation and rotation (θ_z). Comparison of the ground-truthed motion and estimation using the gait-based and the STK-based approaches with random noise added: (a) no additive noise, (b) (0.0,12.0) noise. "-" - ground truth, "o" - STK-based estimates, and "×" - gait-based estimates.

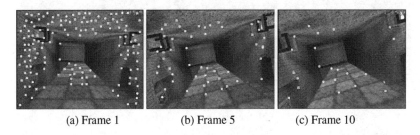

Fig. 9 The "rolling" sequence with the feature points superimposed: (a) frame 1 (150 feature points); (b) frame 5; (c) frame 10 (22 feature points).

In the sequences presented, we defined a forward translation, and the roll or yaw angles of the "observer" were altered according to the expression: roll/yaw angles (deg) = $-6 \times \sin(2\pi \times t_f/30 + 1.1)$, where t_f is the image frame index.

The first image sequence exhibits a pronounced rolling gait. In phase 1, feature tracking shown in Fig. 9 leads to the history of the roll angles shown in Fig. 10(a). This model is used for subsequent tracking with an interval of 3 frames. Fig. 10(b) shows 25 frames immediately succeeding the learning phase, in which gait-based strategy outperforms the STK scheme in terms of measurement accuracy. The longer interval between the neighboring frames leads to the violation of the linearisation assumption in the STK scheme, resulting in large errors in the estimated motion parameters. The second image sequence is of changed yaw angles. Fig. 11 shows feature tracking during the learning phase, leading to the estimated yaw angle (Fig. 11(a)). Similar to the previous experiment, in subsequent frames of which frames 61-120 are shown in Fig. 12 (for yaw angles), the proposed gait method has many fewer measurement errors than the STK framework. Indeed the STK algorithm diverges markedly from the ground truth. Also, in the second cycle (frames 60-120)

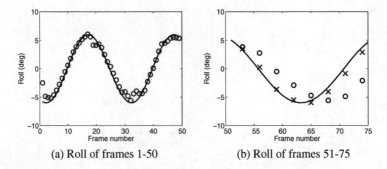

(a) Roll of frames 1-50   (b) Roll of frames 51-75

Fig. 10 Performance comparisons of the STK and the gait-based tracker: the line is the ground truth, the circle the STK, and the cross is the gait-based tracker.

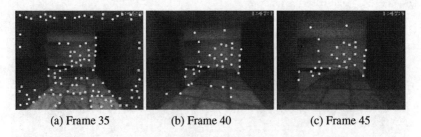

(a) Frame 35   (b) Frame 40   (c) Frame 45

Fig. 11 The "yawing" sequence with the feature points superimposed: (a) frame 35 (57 feature points); (b) frame 40; (c) frame 45 (32 feature points).

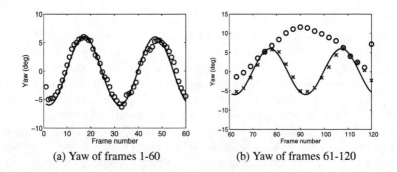

(a) Yaw of frames 1-60   (b) Yaw of frames 61-120

Fig. 12 Performance comparisons of the STK and the gait-based tracker ("yaw" angle estimation): the line is the ground truth, the circle is the STK, and the cross is the gait-based tracker.

the gait based system demonstrates divergence to some degree, due to the significant intensity changes that cause less correct correspondences over frames. Taken overall, these more complex synthetic sequences reinforce the conclusions of the earlier experiments in that the use of gait improves the accuracy of recovery of the ego-motion, and allows more efficient tracking as the time between frames can be increased.

5.4 Real Image Sequences

We acquired several sequences of typical duration 10s from a camera mounted on the waist of a male pedestrian. A photograph of the experimental arrangement can be found in [40]. The images were 360×288 *pixels*^{2} and the frame rate was 25 Hz. The shutter speed was 1/3000 sec. 150 corner features are selected in the first frame of the image sequence. The number of these features decreases gradually as the tracking proceeds. Although the real scene data does not have many occlusions, visual features are lost as the pedestrian view changes, especially when rotating, and features are also lost due to changes in lighting. After ego-motion tracking and recovery of point-wise structure, a technique introduced by Azarbayejiani *et al.* [1] was applied to create texture maps. Vertices from recovered 3-D points were selected manually and then back projected onto assumed planes to yield 3-D polygons. These polygons were combined with the recovered motion and the original video sequence to render texture-mapped models by inverse projecting image pixels onto the scene and matching the pixels with 3-D points on particular objects. These texture maps were compared to the real images for subjective assessment only; no manual intervention was required for the algorithm.

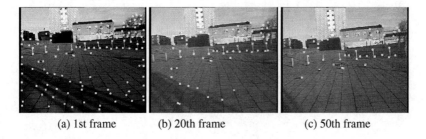

(a) 1st frame   (b) 20th frame   (c) 50th frame

Fig. 13 Video sequence of slow walking with the feature points superimposed

Fig. 13 illustrates some examples from an image sequence collected by the camera as the pedestrian walked slowly, where the detected and tracked feature points are superimposed. This is the first phase, so no gait-model is employed. The number of tracked features decreases as some disappear from view, and others are lost due to incorrect correspondence. Figs. 14 (a)-(f) show the recovered camera positions based on the estimated motion. Many of the detected features are on the ground plane, which can be determined using a robust least median of squares method [43]. Fig. 14 (g) illustrates the saggital view of the 3-D feature positions, where the points ("+") are those included in the ground plane (these points are outlined by two dashed lines), and other points ("⊕") are treated as "objects" as they lie above the ground plane. Using the IRLS approach, we obtain the fitted curves for the camera pose estimates, which are shown in Fig. 15. (t_x and t_y have much smaller variation than t_z, and are omitted.)

We again determined whether use of the gait-based algorithm leads to an improved tracking of features by better predicting their position in future frames (and so losing fewer of them) or, alternatively, whether the algorithm allowed us to sample and process frames at a lower frequency, so improving the efficiency. Fig. 16

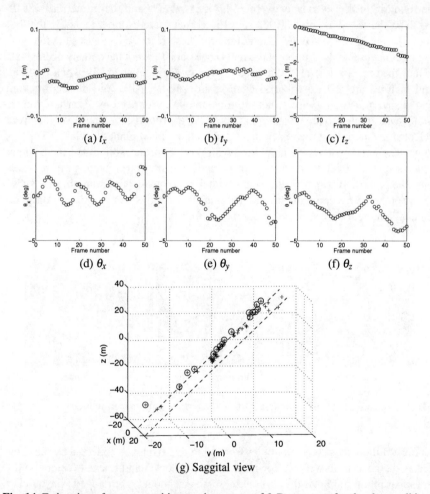

Fig. 14 Estimation of camera positions and recovery of 3-D structure for the slow walking sequence (crosses in (g) indicate the points on the plane while circled crosses refer to obstacles, and hereafter).

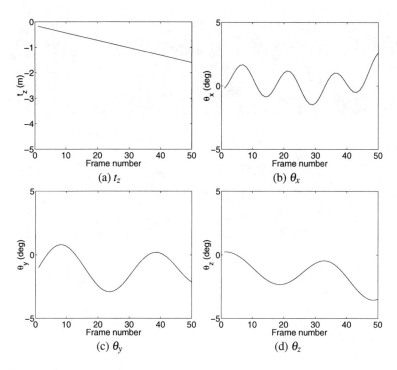

Fig. 15 Fitting of the recovered camera positions for the slow walking sequence

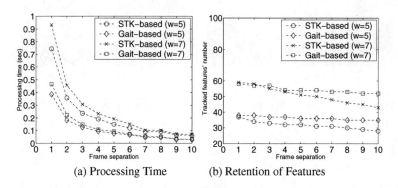

Fig. 16 Comparison between the STK-based and gait-based framework in tracking for the slow walking sequence (w - width of search window)

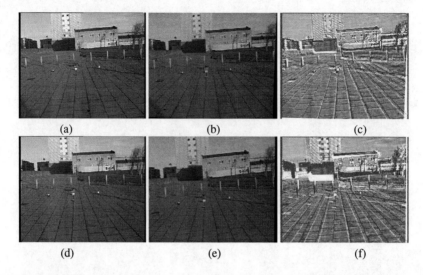

Fig. 17 Comparison of real frames and their texture-maps for the slow walking sequence: (a) The actual frame 55. (b) The texture map of frame 55 created by rendering. (c) Subtraction of (a) and (b). (d) The actual frame 65. (e) The texture map of frame 65 created by rendering. (f) Subtraction of (d) and (e). For better visibility, the brightness/contrast of (c) and (f) has been increased.

Fig. 18 Video sequence of "two pedestrians" with the feature points superimposed

summarizes a performance comparison between the STK- and gait-based frameworks in tracking the sequence of Fig. 13 using a Pentium II-300 MMX PC. As the sampling rate is decreased, the processing time is reduced in an inverse relationship in both cases as there is simply less processing, but the gait-based approach is significantly quicker. Compared to the STK-based approach using a fixed number of pyramid levels, the gait-based method localizes the search better and hence reduces the time taken to find a match. In Fig. 16 (b), the STK-based strategy loses progressively more feature points, and the difference between that and the gait-based prediction is understandably greater as the sampling rate decreases. The prediction that includes the motion model is more robust. To some extent, the data flatters the STK algorithm, first because some of the features lost in each case occur simply

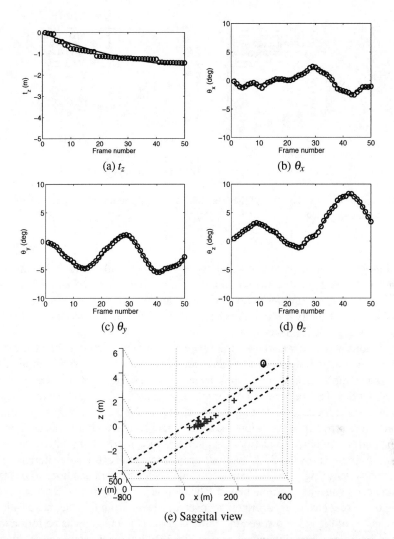

Fig. 19 Estimated camera positions (and their fitted curves) and 3-D structure for the "two pedestrians" sequence. Note that the translations along with x- and y-axes are omitted here due to small variations.

because they leave the field of view of the camera, and second because the oscillatory motion in this sequence is not large. As we do not have ground truth, Fig. 17 shows comparisons between the real (a) and (d), and the texture mapped (b) and (e) scenes for two frames using the estimated motion parameters. The subtraction images (c) and (f) show subjectively the accuracy of the approach.

In the second example, we investigate whether or not moving objects will affect the performance of the proposed motion tracking algorithm. This is a common case in real surveillance scenarios. Figs. 18-20 illustrate the results of the proposed

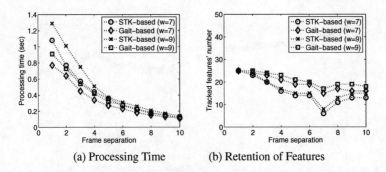

Fig. 20 Comparison between the STK-based and gait-based framework in tracking for the two pedestrians sequence.

algorithm for a video sequence in which two pedestrians walk towards the camera. Similar to the results of the slow walking sequence, the proposed algorithm possesses less feature loss and faster convergence than the STK scheme. This is due to gradual motion of the two moving pedestrians, resulting in very slow feature loss throughout the whole sequence.

An obvious potential weakness of the longer-term motion model is a change in parameters as the pedestrian wearing the camera alters course. Corner turning is a challenge to simple tracking algorithms, but a common occurrence in pedestrian sequences. Tracked feature points may abruptly disappear, leading to unstable motion estimation as the reduced correspondences affect the computation of pose parameters, and more errors in registration due to the appearance of additional candidates from behind the corners. To justify the validity of the proposed gait-based method in this case we first used the STK-based strategy to track the extracted feature points across 100 frames (Figs. 21-23). This period allows the camera positions to be stably estimated in an entire turning period. The motion pattern of the pedestrian when turning the corner has been accurately recovered, as seen from the regular variation of t_x, t_z and θ_y of Fig. 23. However, the gait-based tracker maintains its performance better as the frame separation increases (see Fig. 22). Without a complete (or almost complete) walking history at the corner, one may not obtain a proper gait pattern. With an incorrect prediction, it is unlikely that we will have correct correspondences between frames, leading to the failure of ego-motion tracking. In our case, the success of the gait-based algorithm can also be attributed to the slow postural changes, i.e. the updating of the motion model is sufficiently frequent in relation to the speed of turning the corner. However, if this were not the case, then we acknowledge that the algorithm might not produce fast and correct motion estimates as there would be larger errors in the extracted gait model.

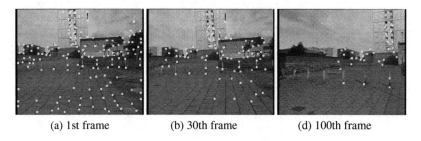

(a) 1st frame (b) 30th frame (d) 100th frame

Fig. 21 Video sequence of walking, turning corner, with the feature points superimposed

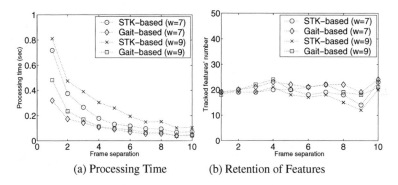

(a) Processing Time (b) Retention of Features

Fig. 22 Comparison between the STK-based and gait-based framework in tracking for the turning corner sequence (w - width of search window).

5.5 Summary of the Experimental Results

We have evaluated the proposed system experimentally. In general, the proposed algorithm is capable of efficiently and effectively tracking feature points in varied circumstances, while providing accurate motion parameters of the camera.

In the case of the synthetic checked targets, Table 1 shows that the gait-based algorithm has RMS errors of 0.15 units less than the STK tracker on average. Fig. 6 indicates that the gait-based algorithm results in errors of less than 0.5 degrees on average in the angular estimates. Changing the sample rates using the synthetic checked targets with changing sampling rates, Fig. 8 demonstrates that the gait-based algorithm allowed optimal recovery of transformations using every sixth frame whereas the STK algorithm requires a frame-by-frame basis.

In the case of synthetic human motion using the games engine, Fig. 10 demonstrates that the proposed gait-based algorithm has errors in roll estimation of 2.0 degrees less than the STK tracker. In the case of real image sequences, although there is no ground truth data, the recovered texture maps show that the proposed algorithm is subjectively better than the STK method (Fig. 17).

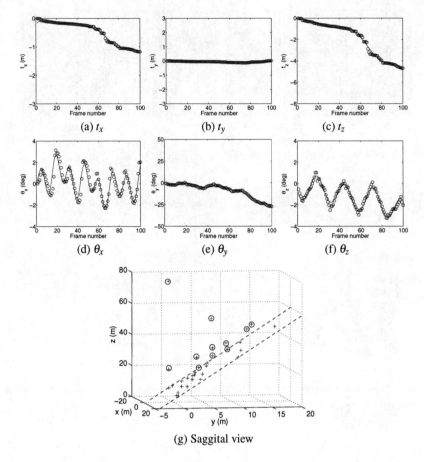

Fig. 23 Estimated camera positions (and their fitted curves) and 3-D structure ("o" - estimates and "-" - fitted curves)

Hence, using both synthetic and real sequences, we have shown that the gait-based method is more efficient than the basic STK tracker in all cases. Furthermore, the gait based approach has allowed greater accuracy of structure and motion parameter recovery in the synthetic sequences, when we have had ground truth available for comparison.

6 Conclusions and Future Work

We have developed a framework for efficient and robust ego-motion tracking using a single camera attached to a pedestrian, i.e. without using any other indicators of position, speed or inclination. An initial gait model is extracted from a fixed training period of two strides, using feature correspondences to estimate the ego-motion parameters, represented as a truncated Fourier series of up to three harmonics for each component of motion. Experiments on synthetic and real data have shown that the proposed strategy has more accurate and efficient ego-motion estimates and structural recovery than a comparable method that does not incorporate long-term motion estimates. The method is robust, and works well in an environment in which the illumination is relatively static. Constantly moving or changing shadows, caused for example by moving bushes or clouds covering the sun, would inevitably cause more features to appear, disappear or shift, and would necessitate an even more robust and complicated algorithm. This is beyond the current scope and would have to be included in future work.

Acknowledgments. We thank Iain Wallace for helping generate the computer game scenarios using the Quake game engine.

References

1. Azarbayejiani, A.J., Galyean, T., Horowitz, B., Pentland, A.: Recursive estimation for CAD model recovery. In: Proc. of the 2nd CAD-Based Vision Workshop, pp. 90–97. IEEE Computer Society, Los Alamitos (1994)
2. Beaton, A., Tukey, J.: The fitting of power series, meaning polynormials, illustrated on band-spectroscopic data. Technometrics 16, 147–185 (1974)
3. Broida, T., Chellappa, R.: Estimation of object motion parameters from noisy images. IEEE Trans. on Pattern Analysis and Machine Intelligence 8, 90–99 (1986)
4. Cappozzo, A.: Analysis of the linear displacement of the head and trunk during walking at different speeds. J. Biomechanics 14, 411–425 (1981)
5. Choi, K., Carcassoni, M., Hancock, E.: Recovering facial pose with the EM algorithm. Pattern Recognition 35(10), 2073–2093 (2002)
6. Coorg, S., Teller, S.: Spherical mosaics with quaternions and dense correlation. International Journal of Computer Vision 37(3), 259–273 (2000)
7. Cross, R., Hancock, E.: Graph matching with a dual-step EM algorithm. IEEE Transactions on Pattern Analysis and Machine Intelligence 20, 1236–1253 (1998)
8. Davison, A., Molton, N., Reid, I., Stasse, O.: MonoSLAM: Real-time single camera slam. IEEE Trans. Pattern Anal. Mach. Intell. 29(6), 1052–1067 (2007)
9. Demidenko, E., Stukel, T.A.: Efficient estimation of general linear mixed effects models. J. of Stat. Plan. and Infer. 104, 197–219 (2002)
10. Dempster, A.: Covariance selection. Biometrics 28, 157–175 (1971)
11. Dempster, A., Laird, N., Rubin, D.: Maximal likelihood from incomplete data via the EM algorithm. J. R. Statist. Soc. B 39, 1–38 (1977)
12. Han, B., Comaniciu, D., Zhu, Y., Davis, L.: Sequential kernel density approximation and its application to real-time visual tracking. IEEE Trans. Pattern Anal. Mach. Intell. 30(7), 1186–1197 (2008)
13. Hartley, R., Zisserman, A.: Multiple View Geometry in Computer Vision, 2nd edn. Cambridge University Press, Cambridge (2004)
14. Jang, D., Choi, H.: Active models for tracking moving objects. Pattern Recognition 33, 1135–1146 (2000)

15. Jones, R., Brelsford, W.: Time series with periodic structure. Biometrika 54(3-4), 403–408 (1992)
16. Laird, N., Ware, J.: Random-effects models for longitudinal data. Biometrics 38, 963–974 (1982)
17. Lepetit, V., Fua, P.: Monocular model-based 3D tracking of rigid objects: a survey. Foundations and Trends in Computer Graphics and Vision 1(1), 1–89 (2005)
18. Meilijson, I.: A fast improvement to the EM algorithm on its own terms. J. R. Statist. Soc. B 51, 127–138 (1989)
19. Molton, N., Brady, J.: Modelling the motion of a sensor attached to a walking person. Robotics and Autonomous Systems 34, 203–221 (2001)
20. Nava, F., Martel, A.: Wavelet modeling of contour deformations in Sobolev spaces for fitting and tracking applications. Pattern Recognition 36, 1119–1130 (2003)
21. Nistér, D.: Preemptive RANSAC for live structure and motion estimation. In: Proc. of the Ninth IEEE International Conference on Computer Vision, p. 199. IEEE Computer Society, Washington (2003)
22. Oliensis, J.: A critique of structure-from-motion algorithms. Computer Vision and Image Understanding 80(2), 172–214 (2000)
23. Press, W., Teukolsky, S., Vetterling, W., Flannery, B.: Numerical Recipes in C: the Art of Scientific Computing, 2nd edn. Cambridge University Press, Cambridge (1992)
24. Rao, C.: The theory of least squares when the parameters are stochastic and its application to the analysis of growth curves. Biometrika 52, 447–458 (1965)
25. Ristic, B., Arulampam, S., Gordon, N.: Beyond the Kalman Filter: Particle Filters for Tracking Applications. Artech House, Boston (2004)
26. Schindler, K., U, J., Wang, H.: Perspective n-view multibody structure-and-motion through model selection. In: Leonardis, A., Bischof, H., Pinz, A. (eds.) ECCV 2006. LNCS, vol. 3951, pp. 606–619. Springer, Heidelberg (2006)
27. Shi, J., Tomasi, C.: Good features to track. In: International Conference on Computer Vision and Pattern Recognition, pp. 593–600 (1994)
28. Smith, S., Brady, J.: SUSAN: A new approach to low-level image-processing. Int. J. of Comput. Vis. 23(1), 45–78 (1997)
29. Snavely, N., Seitz, S., Szeliski, R.: Photo tourism: exploring photo collections in 3D. In: SIGGRAPH, pp. 835–846. Press (2006)
30. Sutherland, D., Olshen, R., Biden, E., Wyatt, M.: The development of mature walking. Blackwell Scientific Publications, Malden (1988)
31. Tao, D., Li, X., Wu, X., Maybank, S.: Elapsed time in human gait recognition: A new approach. In: Proc. of International Conference on Acoustics, Speech, and Signal Processing, France, pp. 177–180 (2006)
32. Tao, D., Li, X., Wu, X., Maybank, S.: General tensor discriminant analysis and Gabor features for gait recognition. IEEE Trans. Pattern Anal. Mach. Intell. 29(10), 1700–1715 (2007)
33. Tomasi, C., Kanade, T.: Shape and motion from image streams under orthography: A factorization method. Int. J. of Comput. Vis. 9(2), 137–154 (1992)
34. Vidal, R., Hartley, R.: Three-view multibody structure from motion. IEEE Trans. Pattern Anal. Mach. Intell. 30(2), 214–227 (2008)
35. Vidal, R., Ma, Y., Hsu, S., Sastry, S.: Optimal motion estimation from multiview normalized epipolar constraint. In: Proc. Int'l Conf. Computer Vision, pp. 34–41 (2001)
36. Xu, G., Zhang, Z.: Epipolar Geometry in Stereo, Motion and Object Recognition: a Unified Approach. Kluwer Academic Publishers, Dordrecht (1996)
37. Xu, Y., Roy-Chowdhury, A.: Integrating motion, illumination, and structure in video sequences with applications in illumination-invariant tracking. IEEE Trans. Pattern Anal. Mach. Intell. 29(5), 793–806 (2007)

38. Zhang, Z.: On the optimization criteria used in two-view motion analysis. IEEE Trans. Pattern Anal. Mach. Intell. 20(7), 717–729 (1998)
39. Zhang, Z., Deriche, R., Faugeras, O., Luong, Q.: A robust technique for matching two uncalibrated images through the recovery of the unknown epipolar geometry. Artificial Intelligence 78, 87–119 (1995)
40. Zhou, H.: Efficient motion tracking and obstacle detection using gait analysis. Ph.D. thesis, Heriot-Watt University, Edinburgh, UK (2005)
41. Zhou, H., Green, P., Wallace, A.: Efficient motion tracking using gait analysis. In: Proc. of International Conference on Acoustics, Speech, and Signal Processing, pp. 601–604 (2004)
42. Zhou, H., Green, P., Wallace, A.: Fundamental matrix estimation using generalized least squares. In: Proc. of International Conference on Visualisation, Imaging, and Image Processing, pp. 79–85 (2004)
43. Zhou, H., Wallace, A.M., Green, P.R.: A multistage filtering technique to detect hazards on the ground plane. Pattern Recognition Letters 24, 1453–1461 (2003)

Spatio-temporal Dynamic Texture Descriptors for Human Motion Recognition

Riccardo Mattivi and Ling Shao

Abstract. In this chapter we apply the Local Binary Pattern on Three Orthogonal Planes (LBP-TOP) descriptor to the field of human action recognition. We modified this spatio-temporal descriptor using LBP and CS-LBP techniques combined with gradient and Gabor images. Moreover, we enhanced its performaces by performing the analysis on more slices located at different time intevals or at different views. A video sequence is described as a collection of spatial-temporal words after the detection of space-time interest points and the description of the area around them. Our contribution has been in the description part, showing LBP-TOP to be 1) a promising descriptor for human action classification purposes and 2) we have developed several modifications and extensions to the descriptor in order to enhance its performance in human motion recognition, showing the method to be computationally efficient.

Keywords: Human Action Recognition, LBP, CS-LBP, LBP-TOP, Bag of Words, Gabor and Gradient images.

1 Introduction

A human action can be defined as an ensemble of movements or behaviours performed by a single person. Automatic categorization and localization of actions in video sequences has different applications, such as detecting activities in surveillance videos, indexing video sequences, organizing digital video library according to specified actions, etc. The challenge for automatically

---

Riccardo Mattivi
Department of Information Engineering and Computer Science,
University of Trento, Italy
e-mail: r.mattivi@disi.unitn.it

Ling Shao
Department of Electronic & Electrical Engineering
The University of Sheffield, Sheffield, S1 3JD, UK
e-mail: ling.shao@sheffield.ac.uk

understanding human actions by computers is to obtain robust action recognition under variable illumination, background changes, camera motion and zooming, viewpoint changes and partial occlusions. Moreover, the system has to cope with a high intraclass variability: the actions can be performed by people wearing different clothes, having different postures and size.

Different approaches in the field of human action recognition have been proposed and developed in the literature: holistic and part-based representations.

Holistic representations focus on the whole human body trying to search characteristics such as contours or pose. Usually holistic methods, which focus on the contours of a person, do not consider the human body as being composed of body parts but consider the whole form of human body in the analyzed frame. Efros et al. [7] use cross-correlation between optical flow descriptors in low resolution videos. However, subjects must be tracked and stabilized and if the background is non uniform, a figure-ground segmentation is required. Bobick et al. [4] use motion history images that capture motion and shape to represent actions. They introduced the global descriptors motion energy image and motion history image. However, their method depends on background subtraction. This method has been extended by Weinland et al. [20]. Shechtman et al. [18] use similarity between space-time volumes which allows finding similar dynamic behaviors and actions, but can not handle large geometric variation between intra-class samples, moving cameras and non-stationary backgrounds.

Motion and trajectories are also commonly used features for recognizing human actions and this could be defined as pose estimation in holisic approaches. Ramanan and Forsyth [15] tracks body parts and then use the obtained motion trajectories to perform action recognition. In particular, they track the humans in the sequences using a structure procedure and then 3D body configurations are estimated and compared to a highly annotated 3D motion library. Multiple cameras and 4D trajectories are used by Yilmaz et al. [21] to recognizing human actions in videos acquired by uncalibrated and moving cameras. They proposed to extend the standard epipolar geometry to the geometry of dynamic scenes and showed the versatility of such method for recognizing of actions in challenging sequences. Ali et al. [3] use trajectories of hands, feet and body. The human body is modelled from experimental data as a nonlinear and chaotic system.

Holistic methods may depend on the recording conditions such as position of the pattern in the frame, spatial resolution, relative motion with respect to the camera and can be influenced by variations in the background and by occlusions. These problems can be solved in principle by external mechanisms (e.g. spatial segmentation, camera stabilization, tracking etc.), but such mechanisms might be unstable in complex situations and require more computational demand.

Part-based representations typically search for Space-Time Interest Points (STIPs) in the video, apply a robust description of the area around them and create a model based on independent features (Bag of Words) or a model that can also contain structural information. These methods do not require tracking and stabilization and are often more resistant to cluttering, as only few parts may be occluded. The resulting features often reflect interesting patterns that can be used for a compact representation of video data as well as for interpretation of spatio-temporal events. Different methods for detecting STIPs have been proposed, such as [10, 6].

This work is an extension of the work done in [12], by comparing several methods with two different kind of learning methods.

The book chapter is organized as follows. In Section 2 we provide the methodology adopted for classification and in Section 3 we provide an introduction to the LBP and LBP-TOP descriptors on 3D data. Experimental results on human action recognition are shown and evaluated in Section 4. Finally, we conclude in Section 5.

2 Methodology

In the following sub-sections we describe our algorithm in detail. In Sub-section 2.1 we explain the classification scheme of our algorithm. In Sub-section 2.2 we describe the detection phase of STIPs, while the feature description phase is breafly introduced in Sub-section 2.3. Sub-section 2.4 explains the classifiers used for training the system.

2.1 Bag of Words Classification

The methodology used in this work is an extension of Bag of Words (BoW) model to video sequences and it has been introduced by Dollar et al. [6]. As a first step, the Space-Time Interest Points, which are the locations where interesting motion is happening, are detected in a video sequence using a separable linear filter. This phase is called Space-Time Interest Points detection. Small video patches (also named as cuboids) are extracted from each STIP. They represent the local information used to learn and train the system for recognizing the different human actions. Each cuboid is then described using the LBP-TOP descriptor. This phase is named as Space-Time Interest Points description.

The result is a sparse representation of the video sequence as cuboid descriptors; the original video sequence can be discarded.

In the training phase, a visual vocabulary (also named as codebook) is built by clustering all the descriptors taken from all training videos. The clustering is done using the k-means algorithm and the center of each cluster is defined as a spatio-temporal 'word' whose length depends on the length of the descriptor adopted. Each feature description is successively assigned to the closest (we use Euclidean distance) vocabulary word and a histogram of spatio-temporal words occurrence is computed for each training videos. Thus, each video is represented as a collection of spatial-temporal words from the codebook in the form of a histogram. The histograms are the data used to train the non-linear Support Vector Machines (SVM) and the k-Nearest Neighbors (kNN) classifier.

In the testing phase, an analogous approach is used: for each video, STIPs are detected and described as feature vectors. Each feature descriptor is successively assigned to the closest visual word stored in the vocabulary. The histogram of spatio-temporal words is then given as input to the classifier whose output is the class of each video.

As the algorithm has a random component, the clustering phase, any experiment result reported is averaged over 20 runs. The entire methodology used is shown in Fig. 1

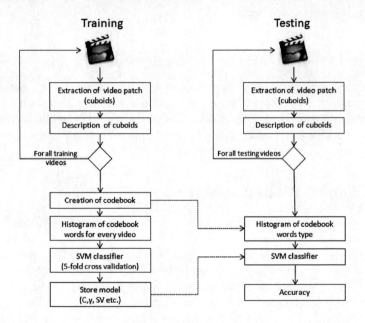

Fig. 1 Methodology

2.2 Feature Detection

Several spatio-temporal feature detection methods have been developed recently and among them we chose Dollar's feature detector [6] because of its simplicity, fastness and because it generally produces a high number of responses. The detector is based on a set of separable linear filters which treats the spatial and temporal dimensions in different ways. A 2D Gaussian kernel is applied only along the spatial dimensions (parameter σ to be set), while a quadrature pair of 1D Gabor filters are applied only temporally (parameter τ to be set). This method responds to local regions which exhibit complex motion patterns, including space-time corners. The response function is given by

$$R = (I * g * h_{even}^2) + (I * g * h_{odd}^2) \qquad (1)$$

where $*$ denotes the convolution operation, $G(x,y,\sigma)$ is the 2D Gaussian kernel, applied only along the spatial dimensions, and h_{even} and h_{odd} are a quadrature pair of the 1D Gabor filters applied only temporally. They are defined as $h_{even}(t;\tau,\sigma) = -\cos(2\pi t\omega)\exp(-t^2/\tau^2)$ and $h_{odd}(t;\tau,\sigma) = -\sin(2\pi t\omega)\exp(-t^2/\tau^2)$; $\omega = 4/\tau$ as suggested by the author.

For more implementation details, please refer to [6] as the feature detection part is beyond the scope of this chapter.

2.3 Feature Description

Once the cuboid is extracted, it is described using the LBP-TOP descriptor, which is an extension of LBP operator into the temporal domain. LBP has originally been proposed for texture analysis and classification [14]. Recently, it has been applied on face recognition [1] and facial expression recognition [2, 17]. While the original LBP was only designed for static images, LBP-TOP has been used for dynamic textures and facial expression recognition [22]. As a video sequence can not only be seen as the usual stack of XY planes in the temporal axis, but also as a stack of YT planes on X axis and as a stack of XT planes on Y axis, we prove that a cuboid can be successfully described with LBP-TOP for action recognition purposes.

2.4 Classification

Each video sequence is described as a histogram of space-time words occurrence which represents its signature. The dimension of the signature is equal to the size of the codebook and the histogram of each videos is given as input to the classifier (see Fig. 1). We chose to use non linear Support Vector Machines (SVM) with rbf kernel and the library libSVM [?] was adopted. The best parameters C and γ were chosen doing a 5-fold cross validation in a grid approach on the training data and one against one approach has been used for multi-class classification.

3 Feature Description: (CS)LBP-TOP and Its Extensions

The original Local Binary Pattern (LBP) operator was introduced by Ojala et al. [13] and was proved to be a powerful texture descriptor. In the original version, the operator labels the pixels of an image by thresholding a 3×3 neighborhood region of each pixel with the center value and considering the results as a binary number. The resulting 256-bin histogram of the computed

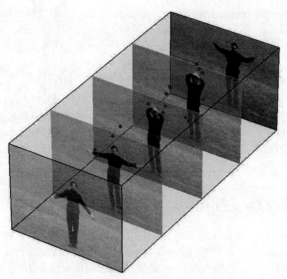

(a) Feature detection

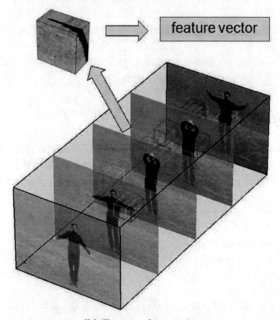

(b) Feature description

Fig. 2 Feature detection and feature description phase

LBP labels is used as a texture descriptor. In Figure 3 the computation of the original LBP is shown. The resulting binary numbers encode local primitives such as curved edges, spots, flat areas etc. as shown in Figure 4.

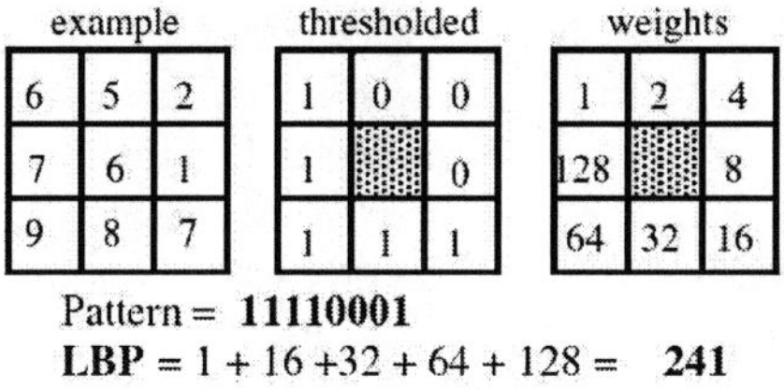

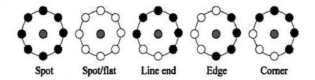

Fig. 3 Example of LBP computation on a 3×3 neighborhood region

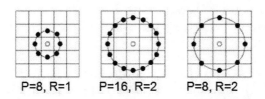

Fig. 4 Local primitives encoded by LBP

Fig. 5 Different values of R and P for LBP

Due to the small neighborhood, the original LBP cannot capture dominant features with large scale structures. The operator was later extended to use neighborhood of different sizes [14]. Circular sampling and bilinear interpolation allow any value for radius and number of pixels in the neighborhood. Given a pixel at a certain location, the resulting LBP operator can be expressed in decimal form as

$$LBP_{P,R} = \sum_{p=0}^{P-1} s(g_p - g_c) 2^P \quad (2)$$

where the notation (P, R) denotes a neighborhood of P points equally sampled on a circle of radius R, g_c is the gray-level value of the central pixel and g_p are the P gray-level values of the sampled points in the neighborhood. $s(x)$ is 1 if $x \geq 0$ and 0 otherwise. The $LBP_{P,R}$ operator produces 2^P different output values, corresponding to the 2^P different binary patterns that can be formed by the P pixels in the neighbor set. It has been shown that certain bins contain more information than others [14]. These patterns are called uniform patterns and are obtained if LBP contains at most two bitwise transitions from 0 to 1 or vice versa when the binary string is considered circular. For example, 00011000, 11000001, 00000000 are uniform patterns. The uniform patterns represent fundamental local primitives, such as edges or corners, as show in Figure 4. It was also observed that most of the texture information (90%) is contained in the uniform patterns [14]. The patterns which have more than two transitions are given a unique label, therefore the operator, denoted $LBP_{P,R}^{u2}$, will have less than 2^P bins. The standard operator $LBP_{8,2}$ will result in 256 different labels while $LBP_{P,R}^{u2}$ will have only 59 labels. After the computation of the LBP labels, a histogram is constructed as follows

$$H_i = \sum_{x,y} I(f_l(x,y) = i), i = 0, 1, ..., L-1 \tag{3}$$

where L is the number of different labels produced by the LBP operator, f_l is the LBP code of the central pixel (x, y) and I(A) is 1 if A is true and 0 otherwise. Moreover, LBP has been extended for multiresolution analysis [14]. The usage of different values for P and R permits to realize operators for any quantization of the angular space and for any spatial resolution. The information provided by multiple operators is then combined. As the LBP histogram contains information about the distribution of local micro-patterns over the whole image, the so computed descriptor represents a statistical description of image characteristics. This descriptor has been proved to be successful, together with its original design as texture description and recognition [14], also in face detection and recognition [1], image retrieval [19] and facial expression analysis and recognition [17].

Another version of LBP has been recently developed by Heikkila et al. [8]: the pixels are compared in a different manner, giving a descriptor whose length is 16 times shorter than LBP. The CS-LBP operator is expressed in decimal form as

$$CS - LBP_{P,R} = \sum_{p=0}^{(P/2)-1} s(g_p - g_{p+(P/2)})2^p \tag{4}$$

In their work, CS-LBP descriptor is proposed to be a powerful descriptor as it combines the strengths of the SIFT descriptor and the LBP texture operator: the method uses a SIFT-like grid approach and replaces SIFT's gradient features with a LBP-based feature. The CS-LBP feature has a relatively short feature histogram, is tolerance to illumination changes and is computational simple. The performance of the CS-LBP descriptor was compared to that of the SIFT descriptor in the contexts of matching and object category classification [8]. For many tests, the proposed CS-LBP method outperforms the SIFT descriptor, while in the other cases, the performance is comparable to SIFT [8].

Recently, LBP has been recently modified in order to be used in the context of dynamic texture description and recognition with also an application to facial expression analysis by Zhao et al. [22]. They introduced the extension of the LBP operator into the temporal domain, named Volume Local Binary Pattern (VLBP), and a different and simplified version which consider the co-occurrences on Three Orthogonal Planes named LBP-TOP. The basic VLBP labels a volume thresholding a neighborhood region not only in the current frame but also in previous and following frames and encoding the results as a binary number. The drawback of this method is that a large number of neighborhood P produces a very long feature vector (2^{3P+2}) while a small P means losing information. LBP-TOP makes the approach computationally simpler and easier extracting the LBP code from three orthogonal planes (XY, XT and YT) denoted as XY-LBP, XT-LBP and YT-LBP. In such a scheme, LBP encodes appearance and motion in three directions, incorporating spatial information in XY-LBP and spatial temporal co-occurrence statistics in XT-LBP and YT-LBP.

LBP-TOP computes the LBP from Three Orthogonal Planes, denoted as XY-LBP, XT-LBP and YT-LBP. The operator is expressed as

$$LBP - TOP_{P_{XY}, P_{XT}, P_{YT}, R_X, R_Y, R_T} \qquad (5)$$

where the notation $P_{XY}, P_{XT}, P_{YT}, R_X, R_Y, R_T$ denotes a neighborhood of P points equally sampled on a circle of radius R on XY, XT and YT planes respectively. The statistics on the three different planes are computed and then concatenated into a single histogram as

$$H_i = \sum_{x,y,t} I(f_j(x,y,t) = i), \qquad i = 0, 1, ..., n_j; \qquad j = 1, 2, 3 \qquad (6)$$

where n_j is the number of different labels produced by the LBP operator in the jth plane, fj is the central pixel at coordinates (x, y, t) in the jth plane and $I(A)$ is 1 if A is true and 0 otherwise. The resulting feature vector is of $3 \cdot 2^P$ length. Fig.6 illustrates the construction of the LBP-TOP descriptor.((In such a scheme, LBP encodes appearance and motion in three directions, incorporating spatial information in XY-LBP and spatial temporal co-occurrence statistics in XT-LBP and YT-LBP)).

In our implementation, LBP-TOP is applied on each cuboid and XY, XT and YT planes are the central slices of it, as shown in Figure 6 and in Figure 7, in which the cuboid is extracted from the running sequence. The

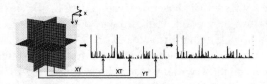

Fig. 6 LBP-TOP methodology

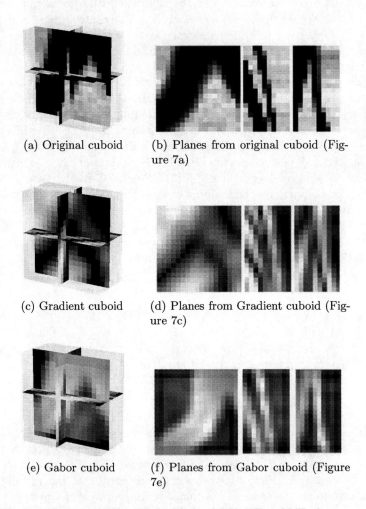

(a) Original cuboid

(b) Planes from original cuboid (Figure 7a)

(c) Gradient cuboid

(d) Planes from Gradient cuboid (Figure 7c)

(e) Gabor cuboid

(f) Planes from Gabor cuboid (Figure 7e)

Fig. 7 Different cuboids with highlighted XY, XT and YT planes

space time interest point detected in the previous feature detection phase is located at the central point in all 3 image planes. In Figure 7 the three extracted planes can be seen and LBP will be applied to each of them. CSLBP-TOP is the computation of CS-LBP operator on Three Orthogonal Planes. Kellokumpu et al [9] have recently used LBP-TOP for human detection and activity description. However, their approach is based on background subtraction using LBP-TOP and a bounding volume has to be built around the area of motion. Their method can be categorize as holistic, since no space-time interest points have to be detected and differs from our part-based approach.

3.1 Modifications on LBP-TOP

As we described previously, the original LBP-TOP descriptor is the computation of LBP on the gray-level values of 3 orthogonal slices of each cuboid (XY, XT and YT planes) resulting in 3 LBP codes (XY-LBP, XT-LBP and YT-LBP).

Extended (CS)LBP-TOP

We propose to extend the computation of LBP to 9 slices, 3 for each axis. Therefore, on the XY dimension we have the original XY plane (centered in the middle of the cuboid) plus other two XY planes located at 1/4 and 3/4 of the cuboid's length. The same is done for XT and YT dimensions. We named this method as Extended LBP-TOP.

In this manner, more dynamic information in the cuboid can be extracted, as the 3 slices in one axis capture the motion at different times. We also exploit more information from the cuboid, dealing with 6 slices on each axis, located from 2/8 until 7/8 of the cuboid's length for each axis. In this case, a dimensionality reduction technique has to be applied since the final dimension of the descriptor vector would be too high.

If the CS-LBP operator is computed on each slice, we define the mehods as Extended CSLBP-TOP.

Gradient and Gabor (CS)LBP-TOP

Another modification we introduced is the computation of LBP operator on gradient or Gabor images of the orthogonal slices. The gradient image contains information about the rapidity of pixel intensity changes along a specific direction, while the gabor image better highlight the area of motion. The gradient of an image, has large magnitude values at edges and it can further increment LBP operator's performances, since LBP encodes local primitives such as curved edges, spots, flat areas etc.

For each cuboid, the brightness gradient is calculated along x, y and t directions, and the resulting 3 cuboids containing specific gradient information are summed in absolute values. Before computing the image gradients, the cuboid is slightly smoothed with a Gaussian filter in order to reduce noise.

LBP-TOP is then performed on the gradient or gabor cuboid and we name this method Gradient LBP-TOP and Gabor LBP-TOP, respectively. In Figure 7 the different cuboids are shown and the three slices XY, XT and YT are highlighted. Figure 10 shows the three XY slices extracted at 1/4, 1/2 and 3/4 of the cuboid's length for the Extended LBP-TOP method.

The Extended LBP-TOP can be applied on the gradient cuboid and we named this method as Extended Gradient LBP-TOP or on the gabor cuboid and we named this method as Extended Gabor LBP-TOP.

If the CS-LBP operator is computed on each slice, we define the mehods as Gradient CSLBP-TOP, Extended Gradient CSLBP-TOP etc. In the result section, more figures are shown.

4 Experimental Results

The dataset we used for our evaluation is KTH human action dataset [16]. This dataset contains six types of human actions: walking, jogging, running, boxing, hand waving and hand clapping. Each action class is performed several times by 25 subjects in different scenarios of outdoor and indoor environment. The camera is not static and the videos present illumination variations, slightly different viewpoint and contain scale changes. In total, the dataset contains 600 sequences. We divide the dataset into two parts: 16 people for training and 9 people for testing, as it has been done in [16] and in [11]. We limit the length of all video sequences to 300 frames.

We extract the STIP and describe the corresponding cuboids with the procedure described in Sections 2.2 and 2.3. The detector parameters of formula 1 are set to $\sigma = 2.8$ and $\tau = 1.6$, which gave better results in our evaluations, and 80 STIPs were detected for each sequence.

Once a cuboid is extracted from the original video sequence, it is described with the proposed methods in order to be robust and discriminative. In the left side of Figure 8 the original video patch extracted from a running sequence is shown, while on the right side each slice of this cuboid is shown side by side. In this Section, we show the results yielded by several methods of feature description.

Fig. 8 Original video patch

In the following figure, the LBP-TOP method is visually explained on real data.

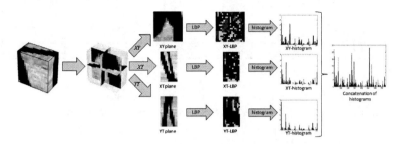

Fig. 9 LBP-TOP methodology

The original LBP-TOP and the Extended LBP-TOP are computed on the original cuboid or on the gradient cuboid. In Table 1 the accuracy results for different values of R and P are shown. The notation of parameters is as illustrated in Equation 5. As it is possible to notice, better classification accuracy has been obtained with the parameter P greater than 6 and radius R equal to 2. The performance is generally slightly decreasing as the radius R is getting bigger, while it is increasing as the number of neighbors P is increased. This could be explained as more neighbors permit to take more information into account, but further distance from the central point means loosing important local information. However, the drawback is a higher computational cost and a higher dimensionality of the feature vector.

$LBP - TOP_{8,8,8,2,2,2}$ produces a 768 vector length, while $LBP - TOP_{10,10,10,2,2,2}$ has a descriptor dimension of 3072. The final descriptor of $LBP - TOP_{12,12,12,2,2,2}$ will be 12288 vector lengths.

Table 1 Accuracy achieved by the SVM classifier for different parameter values of LBP-TOP

		P			
		4	6	8	10
R	2	71.81 %	85.65 %	86.25 %	86.32 %
	3	84.54 %	85.18 %	85.12 %	86.69 %
	4	81.34 %	85.12 %	85.46 %	83.82 %

We have noticed that the use of uniform LBP operator decreases the performance results compared with the original operator, since less information is kept into account (see Table 2). Multiresolution LBP operator has also been tested, but the gain in performances is not considerable with the increase of the descriptor length and computational cost.

Table 2 Accuracy achieved by the SVM classifier for the Original and Uniform LBP-TOP

Method	Accuracy (SVM)	Descriptor length	Computational time (s)
LBP-TOP$_{8,8,8,2,2,2}$	86.25 %	768	0.0139
Uniform LBP-TOP$_{8,8,8,2,2,2}$	81.78 %	177	0.0243

Due to this considerations, we finally choose to use LBP-TOP$_{8,8,8,2,2,2}$ for the following experiments as it is computationally more efficient and the accuracy is among the highest.

The time calculated in the following tables is measured on a computer equipped with a 3 Ghz Pentium 4 CPU and 3 Gb RAM. As dimensionality reduction technique, we used Principal Component Analysis (PCA) and set the final dimension to 100.

In Table 3, the Extended LBP-TOP is evaluated and different number of slices is taken into account. As we can see, the Extended LBP-TOP descriptor performs better than the original one, since more information is taken into consideration at different times in XY planes and at different locations in the XT and YT planes. Although best result is obtained with 6 slices on each axis, the computational time is almost double than the Extended LBP-TOP version with 3 slices; because of this issue, in the following we are computing the Extended version on only 3 slices for each axis.

In Figure 10 the three slices on XY plane from the cuboid of Figure 8 are shown.

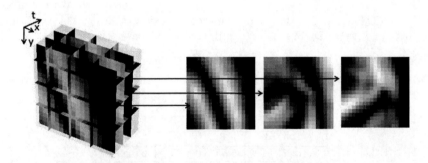

Fig. 10 Extended LBP-TOP: 3 slices in XY plane

In the following tests, the results are showed for different codebook's size, in order to find the best dimension of the codebook, and therefore the feature lenght of the signature of each video, as explained in previous section.

As dimensionality reduction techniques, we applied PCA and OLPP [5, 17], reducing the dimensions to 100. Dealing with a shorter descriptor vector is

Spatio-temporal Dynamic Texture Descriptors

Table 3 Accuracy achieved by the SVM classier for the Original and Extended LBP-TOP

Method	Accuracy (SVM)	Descriptor length	Computational time (s)
LBP-TOP$_{8,8,8,2,2,2}$	86.25 %	768	0.0139
Extended LBP-TOP$_{8,8,8,2,2,2}$ (3 slices on each axis)	88.19 %	2304	0.0314
Extended LBP-TOP$_{8,8,8,2,2,2}$ (3 slices on each axis) +PCA	87.87 %	100	0.0319
Extended LBP-TOP$_{8,8,8,2,2,2}$ (6 slices on each axis) +PCA	88.38 %	100	0.0630

auspicable, since the entire system benefits in computational time and memory resources: the creation of the codebook (k-means algorithm), the histogramming of spatial-temporal words, the training and testing of 1-NN and SVM are faster. Figure 11 shows the performances of Extended LBP-TOP descriptor using the original feature vector (2304 feature length) and using the reduces feature vector with PCA and OLPP (100 feature length). PCA technique is very useful, since the performances are not decreasing. In our tests, OLPP dimensionality reduction technique is not giving good results. As we can see, the performance decreases a lot. Although this technique relies in reflecting the intrinsic geometric structure of the given data space, is supervised and has been proved to be effective in facial expression analysis [17], in our human action recognition task it is not useful.

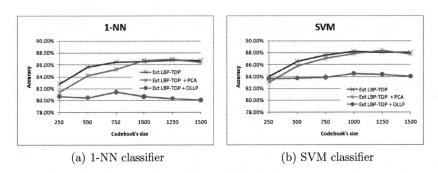

(a) 1-NN classifier   (b) SVM classifier

Fig. 11 Dimensionality reduction techniques (PCA and OLPP) applied to descriptors

Figures 12a and 12b show a comparison for different enhancement of LBP-TOP, as the codebook's size is increasing. The use of LBP-TOP applied to the gradient cuboid gives better results compared with the original one. The information extracted from the gradient calculated along x, y and t

directions and combined into the gradient cuboid permits to have a better performance for LBP-TOP in the description of actions. The gabor filtering applied to the original cuboid helps in increasing the final accuracy for LBP-TOP, although the performance is lower than using the gradient image. This could be explained as the gradient images encode more relevant information for describing the motion inside each video patch than the Gabor images. Moreover, the gradient image better defines the borders of the movement, while gabor image better highlights the area of motion, as shown in Figure 15. A further improvement in the performances can be achieved by applying the Extended LBP-TOP on the gradient or gabor cuboids. The Extended LBP-TOP applied on gabor cuboids is giving very close performance with the Extended LBP-TOP applied on gradient cuboids.

In Figure 13, a comparison between LBP-TOP and CSLBP-TOP is shown, keeping fixed the number of visual words k=1000. As we can notice, the performance of CSLBP-TOP operator is close to that of LBP-TOP, as well as Extended CSLBP-TOP is very similar with Extended LBP-TOP. A higher number of neighbors is needed for CSLBP-TOP to reach better classification accuracy; as the plots show, a number of neighbors equal to 10 or 12 permits to reach performance similar, and even slightly better, to the original LBP-TOP. However, given a fixed number of neighbors, CSLBP-TOP's descriptor is 16 times shorter than that of LBP-TOP. For a number of neighbors equal to 8, the descriptor is 48 length, while LBP-TOP's descriptor is 768 dimensions length.

Figure 14 highlights the best results for LBP-TOP and CSLBP-TOP. The best results have been achieved with the original LBP-TOP implementation. CSLBP operator applied to Gradient or Gabor images gives worse accuracy results.

In general, CSLBP-TOP performs similar with the original LBP-TOP in the field of human action recognition. If the number of neighbors are increased (i.e. $P = 12$), Extended CSLBP-TOP is slightly outperforming the Extended LBP-TOP ($P = 8$), as more spatial information is taken into account during the computation of CS-LBP operator. The Extended Gradient CSLBP-TOP version is performing best among the descriptors based on the CS-LBP operator, reaching almost the performance of Extended Gradient LBP-TOP using 1-NN classifier.

Best performances have been achieved by using the Extended Gradient LBP-TOP. The classification accuracy has been of 92.69% and 92.57% if PCA is applied using the 1-NN classifier with χ^2 distance and setting the codebook's size equal to 1250.

Table 4 shows the computational time for describing one small video patch and classification accuracy. As can be seen, the time is increasing if a higher number of slices is taken into account and if the gradient or gabor cuboid is computed. CSLBP-TOP implementation is slightly faster, as less comparisons have to be computed and the final histogram is shorter. The Extended Gabor LBP-TOP requires more time among all LBP-TOP methods.

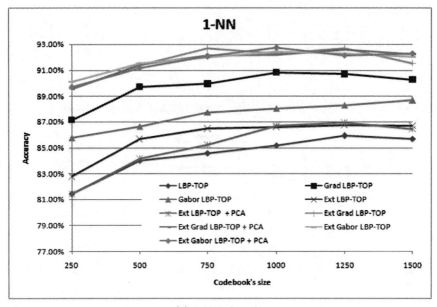

(a) 1-NN classifier

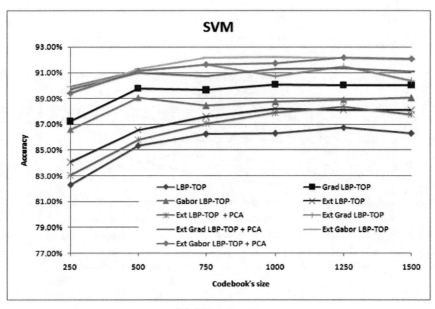

(b) SVM classifier

Fig. 12 Original LBP-TOP and extended LBP-TOP for different enhancements

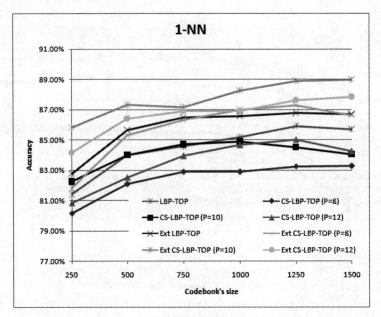

(a) 1-NN classifier

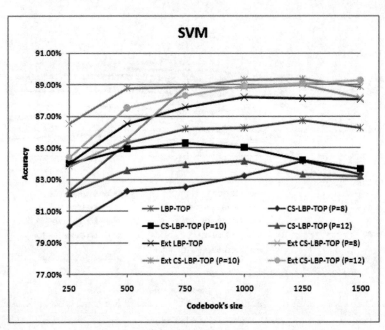

(b) SVM classifier

Fig. 13 LBP-TOP and CSLBP-TOP

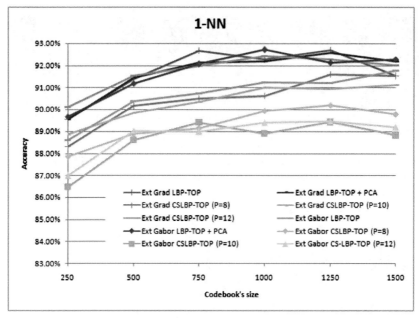

(a) 1-NN classifier

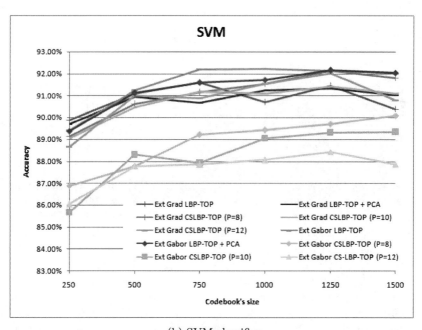

(b) SVM classifier

Fig. 14 Best results for LBP-TOP and CSLBP-TOP

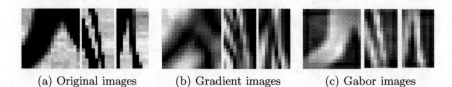

(a) Original images   (b) Gradient images   (c) Gabor images

Fig. 15 Different enhancement of XY, XT and YT images before the computation of LBP operator

Table 4 Computational time and accuracy for different (CS)LBP-TOP implementations, k=1000 visual words

Method	Descriptor length	Computational time (s)	Accuracy (SVM)
LBP-TOP$_{8,8,8,2,2,2}$	768	0.0139	86.25 %
CSLBP-TOP$_{10,10,10,2,2,2}$	96	0.0115	85.00 %
Extended LBP-TOP$_{8,8,8,2,2,2}$	2304	0.0314	88.19 %
Extended LBP-TOP$_{8,8,8,2,2,2}$ + PCA	100	0.0319	87.87 %
Gradient LBP-TOP$_{8,8,8,2,2,2}$	768	0.0788	90.07 %
Extended Gradient LBP-TOP$_{8,8,8,2,2,2}$	2304	0.0992	90.72 %
Extended Gradient LBP-TOP$_{8,8,8,2,2,2}$ + PCA	100	0.1004	91.25 %
Extended Gradient CSLBP-TOP$_{10,10,10,2,2,2}$	288	0.0926	91.09 %
Gabor LBP-TOP$_{8,8,8,2,2,2}$	768	1.037	88.73 %
Extended Gabor LBP-TOP$_{8,8,8,2,2,2}$	2304	1.320	92.22 %
Extended Gabor LBP-TOP$_{8,8,8,2,2,2}$ + PCA	100	1.331	91.71 %

As a comparison, we evaluate Laptev's method [11] with the same framework as illustrated in Section 2 with a codebook size of $k = 1000$ visual words. We use Laptev's code publicly available on his website and recently being updated with the latest settings used in [11]. The combination of Laptev's 3D Harris corner detector and Laptev's HOG-HOF descriptor make us reach an accuracy of 89.88%. In the paper by Laptev [11], 91.8% of accuracy is obtained on KTH database, but different channel combinations for HOG and HOF are being used. Using the terminology of the article, our descriptors are computing in a 1×1 grid. The time for the HOG-HOF descriptor in Table 3 is referred to both detection and description parts, as the description part cannot be computed regardless of the detection part in the provided executable. Therefore, we expect the description part to be about half the time shown in the table. The computational time for HOG-HOF is affected by the choice of the threshold and we have chosen a suitable threshold to have 80 detected STIPs for this comparison. There is also to mention that Laptev's executable code is compiled in C environment, while our LBP-TOP implementation is compiled in Matlab environment. Similar performance to Laptev's is achieved using the Extended LBP-TOP descriptor which is almost 3 times computationally faster than the Extended Gradient LBP-TOP descriptor.

Table 5 Accuracy and computational time for different LBP-TOP methods and HOG-HOF, k=1000 visual words

Method	Descriptor length	Computational time (s)	Accuracy (SVM)
LBP-TOP$_{8,8,8,2,2,2}$	768	0.0139	86.25 %
Ext Grad LBP-TOP$_{8,8,8,2,2,2}$	2304	0.0992	90.72 %
Ext Grad LBP-TOP$_{8,8,8,2,2,2}$ + PCA	100	0.1004	91.25 %
HOG-HOF	162	0.2820*	89.88 %
HOG-HOF + PCA	100	0.2894*	89.28 %

5 Conclusion

In this chapter, we have applied LBP-TOP as a descriptor of small videopatches used in a part-based approach for human action recognition. We have shown that LBP-TOP descriptor can be suitable for the description of cuboids extracted from a video sequences and containing information about human movements and actions.

We have modified the original descriptor introducing the CSLBP-TOP descriptor and we applied the LBP and CS-LBP operator to the original, gradient and Gabor images. Moreover, we extended LBP-TOP and CSLBP-TOP considering the action at three different frames in XY plane and at different views in XT and YT planes. We have also shown that the performance of descriptor is quite stable when the PCA is applied.

The use of Extended Gradient LBP-TOP permits us to reach the best results on the KTH human action database by achieving 92.69% classification accuracy and 92.57% if PCA is applied using 1-NN classifier with χ^2 distance and setting the codebook's size equal to 1250. If SVM classifier is chosen, the classification accuracy is slightly lower, 91.46 % and 91.34% if PCA is applied. In Figure 16 the confusion matrices are shown for both classifiers. Most of the confusion happens between the classes running and jogging, as these actions are very similar to each other, while all actions performed by hands and arms are quite accurately classified.

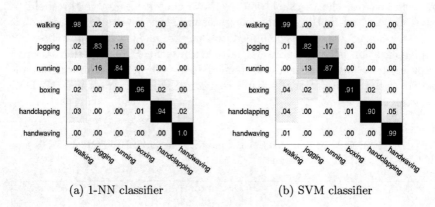

Fig. 16 Confusion matrix for Extended Gradient LBP-TOP + PCA, codebook's size equal to 1250

In addiction, we have also shown the enhanced LBP-TOP descriptor to be more efficient compared with HOG-HOF and permits to reach better accuracy in our framework.

The experimental results reveal that LBP-TOP and its modifications tend to be good candidates for human action description and recognition.

References

1. Ahonen, T., Hadid, A., Pietikainen, M.: Face Recognition with Local Binary Patterns (2004)
2. Ahonen, T., Hadid, A., Pietikainen, M.: Face description with local binary patterns: Application to face recognition. IEEE Transactions on Pattern Analysis and Machine Intelligence 28(12), 2037–2041 (2006)
3. Ali, S., Basharat, A., Shah, M.: Chaotic invariants for human action recognition. In: IEEE International Conference on Computer Vision, vol. 0, pp. 1–8 (2007)
4. Bobick, A.F., Davis, J.W.: The recognition of human movement using temporal templates. IEEE Transactions on Pattern Analysis and Machine Intelligence 23(3), 257–267 (2001)
5. Cai, D., He, X.F., Han, J., Zhang, H.J.: Orthogonal laplacianfaces for face recognition, vol. 15(11), pp. 3608–3614 (2006)
6. Dollar, P., Rabaud, V., Cottrell, G., Belongie, S.: Behavior Recognition via Sparse Spatio-Temporal Features, pp. 65–72 (2005)
7. Efros, A.A., Berg, A.C., Mori, G., Malik, J.: Recognizing Action at a Distance, vol. 2, pp. 726–733 (2003)
8. Heikkilä, M., Pietikäinen, M., Schmid, C.: Description of interest regions with local binary patterns. Pattern Recogn. 42(3), 425–436 (2009)
9. Kellokumpu, V., Zhao, G., Pietikäinen, M.: Human activity recognition using a dynamic texture based method. In: British Machine Vision Conference (2008)
10. Laptev, I., Lindeberg, T.: Space-time interest points. In: ICCV, pp. 432–439 (2003)

11. Laptev, I., Marszalek, M., Schmid, C., Rozenfeld, B.: Learning Realistic Human Actions from Movies, pp. 1–8 (June 2008)
12. Mattivi, R., Shao, L.: Human action recognition using lbp-top as sparse spatio-temporal feature descriptor. In: Jiang, X., Petkov, N. (eds.) CAIP 2009. LNCS, vol. 5702, pp. 740–747. Springer, Heidelberg (2009)
13. Ojala, T., Pietikainen, M., Harwood, D.: A comparative study of texture measures with classification based on feature distributions, vol. 29(1), pp. 51–59 (January 1996)
14. Ojala, T., Pietikainen, M., Maenpaa, T.: Multiresolution gray-scale and rotation invariant texture classification with local binary patterns. IEEE Transactions on Pattern Analysis and Machine Intelligence 24(7), 971–987 (2002)
15. Ramanan, D., Forsyth, D.A.: Automatic annotation of everyday movements. In: NIPS. MIT Press, Cambridge (2003)
16. Schuldt, C., Laptev, I., Caputo, B.: Recognizing human actions: A local svm approach. In: ICPR 2004: Proceedings of the Pattern Recognition, 17th International Conference on (ICPR 2004), Washington, DC, USA, vol. 3, pp. 32–36. IEEE Computer Society, Los Alamitos (2004)
17. Shan, C., Gong, S., McOwan, P.W.: A comprehensive empirical study on linear subspace methods for facial expression analysis. In: CVPRW 2006: Proceedings of the 2006 Conference on Computer Vision and Pattern Recognition Workshop, Washington, DC, USA, p. 153. IEEE Computer Society, Los Alamitos (2006)
18. Shechtman, E., Irani, M.: Space-time behavior based correlation, vol. 1, pp. 405–412 (2005)
19. Takala, V., Ahonen, T., Pietikainen, M.: Block-based methods for image retrieval using local binary patterns, pp. 882–891 (2005)
20. Weinland, D., Ronfard, R., Boyer, E.: Free viewpoint action recognition using motion history volumes (November/December 2006)
21. Yilmaz, A., Shah, M.: Recognizing human actions in videos acquired by uncalibrated moving cameras. In: IEEE International Conference on Computer Vision, vol. 1, pp. 150–157 (2005)
22. Zhao, G., Pietikainen, M.: Dynamic texture recognition using local binary patterns with an application to facial expressions. IEEE Transactions on Pattern Analysis and Machine Intelligence 29(6), 915–928 (2007)

Efficient Object Localization with Variation-Normalized Gaussianized Vectors

Xiaodan Zhuang, Xi Zhou, Mark A. Hasegawa-Johnson, and Thomas S. Huang

Abstract. Effective object localization relies on efficient and effective searching method, and robust image representation and learning method. Recently, the Gaussianized vector representation has been shown effective in several computer vision applications, such as facial age estimation, image scene categorization and video event recognition. However, all these tasks are classification and regression problems based on the whole images. It is not yet explored how this representation can be efficiently applied in the object localization, which reveals the locations and sizes of the objects. In this work, we present an efficient object localization approach for the Gaussianized vector representation, following a branch-and-bound search scheme introduced by Lampert et al. [5]. In particular, we design a quality bound for rectangle sets characterized by the Gaussianized vector representation for fast hierarchical search. This bound can be obtained for any rectangle set in the image, with little extra computational cost, in addition to calculating the Gaussianized vector representation for the whole image. Further, we propose incorporating a normalization approach that suppresses the variation within the object class and the background class. Experiments on a multi-scale car dataset show that the proposed object localization approach based on the Gaussianized vector representation outperforms previous work using the histogram-of-keywords representation. The within-class variation normalization approach further boosts the performance. This chapter is an extended version of our paper at the 1st International Workshop on Interactive Multimedia for Consumer Electronics at ACM Multimedia 2009 [16].

Xiaodan Zhuang · Xi Zhou · Mark A. Hasegawa-Johnson · Thomas S. Huang
Beckman Inst., ECE Dept., UIUC
e-mail: xzhuang2@uiuc.edu, xizhou2@uiuc.edu
  jhasegaw@uiuc.edu, huang@ifp.uiuc.edu

1 Introduction

Object localization predicts the bounding box of a specific object class within the image. Effective object localization relies on efficient and effective searching method, and robust image representation and learning method. The task remains challenging due to within-class variations and the large search space for candidate bounding boxes.

A straightforward way to carry out localization is the sliding window approach [9], which applies learned classifiers over all candidate bounding boxes. However, an exhaustive search in an $n \times n$ image needs to evaluate $O(n^4)$ candidate bounding boxes, and is not affordable with complex classifiers. Tricky heuristics about possible bounding box locations, widths and heights, or local optimization methods would have to be used, resulting in false estimates. Despite the great improvement in computer capabilities, the intrinsic tradeoff between performance and efficiency is not desirable, particularly for applications that are highly efficiency sensitive. In recent years, the most popular technique in the sliding window approach is the cascade [10], which decomposes a strong object/non-object classifier into a series of simpler classifiers arranged in a cascade. However, the cascade is slow to train and involves many empirical decisions. Moreover, it always reduces the performance compared with the original strong classifier. As an alternative to the sliding window approach, Lampert et al. introduced a branch-and-bound search scheme[5], which finds the globally optimal bounding box efficiently without the above problems.

Robust image representation and learning is critical to the success of various computer vision applications. Some of the successful features are Histogram of Oriented Gradients [14] and Haar-like features [10]. Patch-based histogram-of-keywords image representation methods represent an image as an ensemble of local features discretized into a set of keywords. These methods have been successfully applied in object localization [5] and image categorization[3]. The Gaussian mixture model (GMM) is widely used for distribution modeling in speech recognition, speaker identification and computer vision. Recently, the Gaussianized vector representation was proposed as an innovative image and video vector representation based on the GMM [12]. Variants of this Gaussianized vector representation have been successfully applied in several applications related to interactive multimedia, such as facial age estimation [11, 15], image scene categorization [12] and video event recognition [13].

While the Gaussianized vector representation proves effective in the above visual recognition tasks, all these are classification or regression problems working on the whole images. In contrast, the object detection or localization problem finds the rectangle bounding boxes for instances of a particular object with varying locations, widths and heights. However, it is not clear how to use the Gaussianized vector representation to capture localized information besides global information in an image. No work has yet explored applying the Gaussianized vector representation in the object localization problem.

In this work, we present an object localization approach combining the efficient branch-and-bound searching method with the robust Gaussianized vector representation. The branch-and-bound search scheme [5] is adopted to perform fast hierarchical search for the optimal bounding boxes, leveraging a quality bound for rectangle sets. We demonstrate that the quality function based on the Gaussianized vector representation can be written as the sum of contributions from each feature vector in the bounding box. Moreover, a quality bound can be obtained for any rectangle set in the image, with little computational cost, in addition to calculating the Gaussianized vector representation for the whole image.

To achieve improved robustness to variation in the object class and the background, we propose incorporating a normalization approach that suppresses the within-class covariance of the Gaussianized vector representation kernels in the binary Support Vector Machine (SVM) and the branch-and-bound searching scheme.

We carry out object localization experiments on a multi-scale car dataset. The results show the proposed object localization approach based on the Gaussianized vector representation outperforms a similar system using the branch-and-bound search based on the histogram-of-keywords representation. The normalization approach further improves the performance of the object localization system. These suggest that the Gaussianized vector representation can be effective for the localization problem besides the classification and regression problems reported previously.

The rest of this chapter is arranged as follows. In Section 2, we describe the procedure of constructing Gaussianized vector representation. Section 3 presents the normalization approach for robustness to object and background variation. Section 4 details the proposed efficient localization method based on the Gaussianized vector representation. The experimental results on multi-scale car detection are reported in Section 5, followed by conclusions and discussion in Section 6. This chapter is extended from our paper at the 1st International Workshop on Interactive Multimedia for Consumer Electronics at ACM Multimedia 2009 [16].

2 Gaussianized Vector Representation

The Gaussian mixture model (GMM) is widely used in various pattern recognition problems [8, 7]. Recently, the Gaussianized vector representation was proposed. This representation encodes an image as a bag of feature vectors, the distribution of which is described by a GMM. Then a GMM supervector is constructed using the means of the GMM, normalized by the covariance matrices and Gaussian component priors. A GMM-supervector-based kernel is designed to approximate Kullback-Leibler divergence between the GMMs for any two images, and is utilized for supervised discriminative learning using an SVM. Variants of this GMM-based representation have been successfully applied in several visual recognition tasks, such as facial age estimation [11, 15], scene categorization [12] and video event recognition [13].

As pointed out by [12], the success of this representation can be attributed to two properties. First, it establishes correspondence between feature vectors in different images in an unsupervised fashion. Second, it observes the standard normal distribution, and is more informative than the conventional histogram of keywords.

The Gaussianized vector representation is closely connected to the classic histogram of keywords representation. In the traditional histogram representation, the keywords are chosen by the k-means algorithm on all the features. Each feature is distributed to a particular bin based on its distance to the cluster centroids. The histogram representation obtains rough alignment between features vectors by assigning each to one of the histogram bins. Such a representation provides a natural similarity measure between two images based on the difference between the corresponding histograms. However, the histogram representation has some intrinsic limitations. In particular, it is sensitive to feature outliers, the choice of bins, and the noise level in the data. Besides, encoding high-dimensional feature vectors by a relatively small codebook results in large quantization errors and loss of discriminability.

Gaussianized vector representation enhances the histogram representation in the following ways. First, k-means clustering leverages the Euclidean distance, while the GMM leverages the Mahamalobis distance by means of the component posteriors. Second, k-means clustering assigns one single keyword to each feature vector, while the Guassinized vector representation allows each feature vector to contribute to multiple Gaussian components statistically. Third, histogram-of-keywords only uses the number of feature vectors assigned to the histogram bins, while the Gaussianized vector representation also engages the weighted mean of the features in each component, leading to a more informative representation.

2.1 GMM for Feature Vector Distribution

We estimate a GMM for the distribution of all patch-based feature vectors in an image. The estimated GMM is a compact description of the single image, less prone to noise compared with the feature vectors. Yet, with increasing number of Gaussian components, the GMM can be arbitrarily accurate in describing the underlying feature vector distribution. The Gaussian components impose an implicit multi-mode structure of the feature vector distribution in the image. When the GMMs for different images are adapted from the same global GMM, the corresponding Gaussian components imply certain correspondence.

In particular, we obtain one GMM for each image in the following way.

First, a global GMM is estimated using patch-based feature vectors extracted from all training images, regardless of their labels. Here we denote z as a feature vector, whose distribution is modeled by a GMM, a weighted linear combination of K unimodal Gaussian components,

$$p(z;\Theta) = \sum_{k=1}^{K} w_k \mathcal{N}(z; \mu_k^{global}, \Sigma_k).$$

$\Theta = \{w_1, \mu_1^{global}, \Sigma_1, \cdots\}$, w_k, μ_k and Σ_k are the weight, mean, and covariance matrix of the kth Gaussian component,

$$\mathcal{N}(z;\mu_k,\Sigma_k) = \frac{1}{(2\pi)^{\frac{d}{2}}|\Sigma_k|^{\frac{1}{2}}} e^{-\frac{1}{2}(z-\mu_k)^T \Sigma_k^{-1}(z-\mu_k)}. \tag{1}$$

We restrict the covariance matrices Σ_k to be diagonal [8], which proves to be effective and computationally economical.

Second, an image-specific GMM is adapted from the global GMM, using the feature vectors in the particular image. This is preferred to direct seperate estimation of image-specific GMMs for the following reasons. 1) It improves robust parameter estimation of the image specialized GMM, using the comparatively small number of feature vectors in the single image. 2) The global GMM learnt from all training images may provide useful information for the image specialized GMM. 3) As mentioned earlier, it establishes correspondence between Gaussian components in different images-specifc GMMs. For robust estimation, we only adapt the mean vectors of the global GMM and retain the mixture weights and covariance matrices. In particular, we adapt an image-specific GMM by *Maximum a Posteriori* (MAP) with the weighting all on the adaptation data. The posterior probabilities and the updated means are estimated as

$$Pr(k|z_j) = \frac{w_k \mathcal{N}(z_j;\mu_k^{global},\Sigma_k)}{\sum_{k=1}^{K} w_k \mathcal{N}(z_j;\mu_k^{global},\Sigma_k)}, \tag{2}$$

$$\mu_k = \frac{1}{n_k} \sum_{j=1}^{H} Pr(k|z_j) z_j, \tag{3}$$

where n_k is a normalizing term,

$$n_k = \sum_{j=1}^{H} Pr(k|z_j), \tag{4}$$

and $Z = \{z_1,\ldots,z_H\}$ are the feature vectors extracted from the particular image.

As shown in Equation 2, the image-specific GMMs leverage statistical membership of each feature vector among multiple Gaussian components. This sets the Gaussianized vector representation apart from the histogram of keyword representation which originally requires hard membership in one keyword for each feature vector. In addition, Equation 3 shows that the Gaussianized vector representation encodes additional information about the feature vectors statistically assigned to each Gaussian component, via the means of the components.

Given the computational cost concern for many applications, another advantage of using GMM to model feature vector distribution is that efficient approximation exists for GMM that does not significantly degrade its effectiveness. For example, we can prune out Gaussian components with very low weights in the adapted image-specific GMMs. Another possibility is to eliminate the additions in Equation 3 that involves very low priors in Equation 2. Neither of these approaches significantly degrades GMM's capability to approximate a distribution [8].

2.2 Discriminative Learning

Suppose we have two images whose ensembles of feature vectors, Z_a and Z_b, are modeled by two adapted GMMs according to Section 2.1, denoted as g_a and g_b. A natural similarity measure is the approximated Kullback-Leibler divergence

$$D(g_a||g_b) \leq \sum_{k=1}^{K} w_k D(\mathcal{N}(z;\mu_k^a,\Sigma_k)||\mathcal{N}(z;\mu_k^b,\Sigma_k)), \qquad (5)$$

where μ_k^a denotes the adapted mean of the kth component from the image-specific GMM g_a, and likewise for μ_k^b. The right side of the above inequality is equal to

$$d(Z_a,Z_b) = \frac{1}{2}\sum_{k=1}^{K} w_k(\mu_k^a - \mu_k^b)^T \Sigma_k^{-1}(\mu_k^a - \mu_k^b). \qquad (6)$$

$d(Z_a,Z_b)^{\frac{1}{2}}$ can be considered as the Euclidean distance in another high-dimensional feature space,

$$d(Z_a,Z_b) = \|\phi(Z_a) - \phi(Z_b)\|^2$$
$$\phi(Z_a) = [\sqrt{\frac{w_1}{2}}\Sigma_1^{-\frac{1}{2}}\mu_1^a;\cdots;\sqrt{\frac{w_K}{2}}\Sigma_K^{-\frac{1}{2}}\mu_K^a]. \qquad (7)$$

Thus, we obtain the corresponding kernel function

$$k(Z_a,Z_b) = \phi(Z_a) \bullet \phi(Z_b). \qquad (8)$$

A Support Vector Machine (SVM) is used with the above kernel to distinguish objects from backgrounds. The classification score for a test image is

$$g(Z) = \sum_{t} \alpha_t k(Z,Z_t) - b, \qquad (9)$$

where α_t is the learnt weight of the training sample Z_t and b is a threshold parameter. $k(Z,Z_i)$ is the value of a kernel function for the training Gaussianized vector representation Z_i and the test Gaussianized vector representation Z. y_i is the class label of Z_i used in training, indicating whether or not the concerned object is present.

The support vectors and their corresponding weights are learned using the standard quadratic programming optimization process.

3 Robustness to Within-Class Variation

The variation of the object class and the background adds to the difficulty of the localization problem. The Gaussianized vector representation is based on Gaussian mixtures adapted from the global model. To further enhance the discriminating power between objects and the background, we propose incorporating a normalization approach, which depresses the kernel components with high-variation within each class. This method was first proposed in the speaker recognition problem [4], and we have successfully applied it in video categorization [13].

We assume the Gaussianized vector representation kernels in Equation 8 are characterized by a subspace spanned by the projection matrix V^{all}. The desired normalization suppresses the subspace, V, that has the maximum inter-image distance d_V for images (or image regions) of either the objects or the backgrounds:

$$d_V^{ab} = \|V^T \phi(Z_a) - V^T \phi(Z_b)\|^2. \tag{10}$$

Since V identifies the subspace in which feature similarity and label similarity are most out of sync, this subspace can be suppressed by calculating the kernel function as in Equation 11, where C is a diagonal matrix, indicating the extent of such asynchrony for each dimension in the subspace.

$$k(Z_a, Z_b) = \phi(Z_a)^T (I - VCV^T) \phi(Z_b). \tag{11}$$

We can find the subspace V by solving the following,

$$V = \arg\max_{V^T V = I} \sum_{a \neq b} d_V^{ab} W_{ab}, \tag{12}$$

where $W_{ab}=1$ when Z_a and Z_b both belong to the object class or the background class, otherwise $W_{ab} = 0$.

Denote $\hat{Z} = [\phi(Z_1), \phi(Z_2), \cdots, \phi(Z_N)]$, where N is the total number of training images, it can be shown that the optimal V consists of the eigenvectors corresponding to the largest eigenvalues Λ of the matrix $\hat{Z}(D-W)\hat{Z}^T$, where D is a diagonal matrix with $D_{ii} = \sum_{j=1}^{N} W_{ij}, \forall i$.

The eigenvalues Λ indicate the extent to which the corresponding dimensions vary within the same class. In order to ensure the diagonal elements of C remain in the range of $[0,1]$, we apply a monotonic mapping $C = 1 - max(I, \Lambda)^{-1}$.

4 Localization with Gaussianized Vector Representation

In this section, we first present the efficient search scheme based on branch-and-bound in Subsection 4.1. Then we detail the quality function and qualify bound for the Gaussianized vector representation in Subsections 4.2 and 4.3 respectively. In Subsection 4.4, we present incorporating the variation-normalization approach in the localization framework.

4.1 Branch-and-Bound Search

Localization of an object is essentially to find the subarea in the image on which a quality function f achieves its maximum, over all possible subareas. One way to define these subareas is the bounding box, which encodes the location, width and height of an object with four parameters, i.e., the top, bottom, left and right coordinates (t,b,l,r).

The sliding window approach is most widely used in object localization with bounding boxes [9, 2]. To find the bounding box where the quality function f reaches its global maximum, we need evaluate the function on all possible rectangles in the image, whose number is on the order of $O(n^4)$ for an $n \times n$ image. To reduce the computational cost, usually only rectangles at a coarse location grid and of a small number of possible widths and heights are considered. On the other hand, different approaches can be adopted to use a local optimum to approximate the global one, when the quality function f has certain properties, such as smoothness. All these approaches make detection tractable at the risk of missing the global optimum, and with demand for well informed heuristics about the possible location and sizes of the object.

In recent years, the most popular technique in the sliding window approach is the cascade [10]. The cascade technique decomposes a strong object/non-object classifier into a series of simpler classifiers. These classifiers are arranged in a cascade, so that the simpler and weaker classifiers will eliminate most of the candidate bounding boxes, before the more powerful and complicated classifiers will make finer selection. However, the cascade of classifiers is slow to train. Moreover, it unfortunately involves many empirical decisions, e.g., choosing the false alarm rate and missing rate at each stage of the cascade. The cascade technique always reduces the performance compared with the original strong classifier.

The branch-and-bound search scheme was recently introduced [5] to find the globally optimal bounding box without the heuristics and assumptions about the property of the quality function. It hierarchically splits the parameter space of all the rectangles in an image, and discards large parts if their upper bounds fall lower than an examined rectangle.

For localization based on bounding boxes, a set of rectangles is encoded with $[T,B,L,R]$, each indicating a continual interval for the corresponding parameter in (t,b,l,r). The approach starts with a rectangle set containing all the rectangles in the image, and terminates when one rectangle is found that has a quality function no worse than the bounds $\hat{f}$ of any other rectangle set.

At every iteration, the parameter space $[T,B,L,R]$ is split along the largest of the four dimension, resulting in two rectangle sets both pushed into a queue together with their upper bounds. The rectangle set with the highest upper bound is retrieved from the queue for the next iteration.

The steps of the branch-and-bound search scheme can be summarized as follows:

1. Initialize an empty queue Q of rectangle sets. Initialize a rectangle set **R** to be all the rectangles: T and B are both set to be the complete span from zero to the height of the image. L and R are both set to be the complete span from zero to the width of the image.
2. Obtain two rectangle sets by splitting the parameter space $[T,B,L,R]$ along the largest of the four dimension.
3. Push the two rectangle sets in Step 2 into queue Q with their respective quality bound.

Efficient Object Localization with Variation-Normalized Gaussianized Vectors 101

4. Update R with the rectangle set with the highest quality bound in Q.
5. Stop and return R if **R** contains only one rectangle R. Otherwise go to Step 2.

The quality bound $\hat{f}$ for a rectangle set **R** should satisfy the following conditions:

1. $\hat{f}(\mathbf{R}) \geq max_{R \in \mathbf{R}} f(R)$

2. $\hat{f}(\mathbf{R}) = f(R)$, if R is the only element in **R**

Critical for the branch-and-bound scheme is to find the quality bound $\hat{f}$. Given the proven performance of the Gaussianized vector representation in classification tasks shown in previous work [11, 13, 15, 12], we are motivated to design a quality bound based on this representation, to enable localization based on this representation.

4.2 Quality Function

For the Gaussianized vector representation, the binary classification score in Equation 9 informs the confidence that the evaluated image subarea contains the object instead of pure background. Therefore, we can use this score as the quality function for the Gaussianized vector representation.

In particular, according to Equation 8 and Equation 9, the quality function f can be defined as follows,

$$f(Z) = g(Z) = \sum_t \alpha_t \phi(Z) \bullet \phi(Z_t) - b, \qquad (13)$$

which can be expanded using Equation 7,

$$f(Z) = \sum_t \alpha_t \sum_{k=1}^{K} \sqrt{\frac{w_c}{2}} \Sigma_c^{-\frac{1}{2}} \mu_k$$
$$\bullet \sqrt{\frac{w_k}{2}} \Sigma_c^{-\frac{1}{2}} \mu_k^i - b$$
$$= \sum_t \alpha_t \sum_{k=1}^{K} \frac{w_k}{2} \Sigma_k^{-1} \mu_k \bullet \mu_k^t - b. \qquad (14)$$

According to Equation 3, the adapted mean of an image-specifc GMM is the sum of the feature vectors in the image, weighted by the corresponding posterior. Therefore,

$$f(Z) = \sum_t \alpha_t \sum_{k=1}^{K} \frac{w_k}{2} \Sigma_k^{-1} \frac{1}{n_k} \sum_{j=1}^{H} Pr(k|z_j) z_j \bullet \mu_k^t - b.$$
$$= \sum_{j=1}^{H} \left\{ \sum_{k=1}^{K} \frac{1}{n_k} Pr(k|z_j) z_j \bullet \frac{w_k}{2} \Sigma_k^{-1} \sum_t \alpha_t \mu_k^t \right\} - b. \qquad (15)$$

4.3 Quality Bound

We define the "per feature vector contribution" as the contribution of each feature vector in a subarea to the confidence that this subarea is the concerned object. In particular, the "per feature vector contribution" is defined as in Equation 16.

$$W_j = \sum_{k=1}^{K} \frac{1}{n_k} Pr(k|z_j) z_j \bullet \frac{w_k}{2} \Sigma_k^{-1} \sum_t \alpha_t \mu_k^t. \tag{16}$$

Therefore, Equation 15 can be rewritten as Equation 17, showing that the quality function can be viewed as the sum of contribution from all involved feature vectors.

$$f(Z) = \sum_j W_j - b. \tag{17}$$

Given a test image, if we approximate the term n_k with their values calculated on the whole image, the per feature vector contributions $W_j, j \in 1,...,H$ are independent from the bounding box within the test image. This means that we can precompute W_j and evaluate the quality function on different rectangles by summing up those W_j that fall into the concerned rectangle.

We design a quality bound for the Gaussianized vector representation in a way similar to the quality bound for histogram of keywords proposed in [5]. For a set of rectangles, the quality bound is the sum of all positive contributions from the feature vectors in the largest rectangle and all negative contributions from the feature vectors in the smallest rectangle. This can be formulated as

$$\hat{f}(\mathbf{R}) = \sum_{W_{j_1} \in R_{max}} W_{j_1} \times (W_{j_1} > 0) \\ + \sum_{W_{j_2} \in R_{min}} W_{j_2} \times (W_{j_2} < 0). \tag{18}$$

where $[T,B,L,R]$ are the intervals of t,b,l,r and R_{max} and R_{min} are the largest and the smallest rectangles.

We demonstrate that Equation 18 satisfies the conditions of a qualify bound for the branch-and-bound search scheme defined in Section 4.1.

First, the proposed $\hat{f}(\mathbf{R})$ is an upper bound for all rectangles in the set $\mathbf{R}$. In particular, the qualify function evaluated on any rectangle R can be written as the sum of postive contributions and negative contributions from feature vectors in this rectangle,

$$f(R) = \sum_{W_{j_1} \in R} W_{j_1} \times (W_{j_1} > 0) \\ + \sum_{W_{j_2} \in R} W_{j_2} \times (W_{j_2} < 0). \tag{19}$$

Obviously, given a rectangle set **R**, the first term in Equation 19 is maximized by taking all the positive contributions from the largest rectangle in the set. The second term in Equation 19 is negative and its absolute value can be minimized by taking all the negative contributions in the smallest rectangle.

Second, when the rectangle set **R** contains only one rectangle, $R_{min} = R_{max} = R$. Equation 18 equals Equation 19,

$$\hat{f}(\mathbf{R}) = f(R).$$

This quality bound defined by Equation 18 is used in the branch-and-bound scheme discussed in Section 4.1 to achieve fast and effective detection and localization. Note that since the bound is based on sum of per feature vector contributions, the approach can be repeated to find multiple bounding boxes in an image, after removing those features claimed by the previously found boxes. This avoids the problem of finding multiple non-optimal boxes near a previously found box as in the sliding window approach.

Note that estimating W_j in Equation 16 involves no more computation than the calculation in a binary classifier using the Gaussianized vector representation of the whole image. To further expedite the localization, we can use two integral images [10] to speed up the two summations in Equation 18 respectively. This makes the calculation of $\hat{f}(\mathbf{R})$ independent from the number of rectangles in the set **R**.

4.4 Incorporating Variation-Normalization

To further improve the discriminating power of the Gaussianized vector representation in the localization problem, we incorporate the normalization approach in Section 3. In particular, this involves the following modifications of the proposed efficient localization system.

First, the SVM is trained using kernels with normalization against within-class variation. In particular, Equation 11 is used instead of Equation 8.

Second, Equation 13 is replaced by Equation 20 to suppress the subspace that corresponds to the most within-class variation when evaluating the quality of the candidate regions.

$$f(Z) = g(Z) = \sum_{t} \alpha_t \phi(Z)^T (I - VCV^T) \phi(Z_t) - b. \tag{20}$$

Third, the per feature vector contribution function in Equation 16 needs to be revised accordingly.

Let's denote

$$P = \begin{bmatrix} \sqrt{\frac{w_1}{2}}\Sigma_1^{-1/2} & & 0 \\ & \ddots & \\ 0 & & \sqrt{\frac{w_K}{2}}\Sigma_K^{-1/2} \end{bmatrix} \quad (21)$$

$$\quad (22)$$

$$H^t = [H_1^t; \cdots ; H_K^t] \quad (23)$$

$$= P(I - VCV^T)\phi(Z_t), \quad (24)$$

where H^t summarizes information from the tth training image.

With Equations 17, 20 and 21, it can be shown that the per feature vector contribution function can be written as in Equation 25.

$$W_j = \sum_{k=1}^{K}\sum_{t} \alpha_t H_k^t \bullet \frac{1}{n_k} Pr(k|z_j)z_j. \quad (25)$$

5 Experiments

In this paper, we carry out object localization experiments using the proposed efficient object localization approach based on the Gaussinized vector representation. We compare the detection performance with a similar object localization system based on the generic histogram of keywords. In addition, we demonstrate that the proposed normalizing approach can be effectively incorporated in object localization based on Gaussianized vector representation.

5.1 Dataset

We use a multi-scale car dataset[1] for the localization experiment. There are 1050 training images of fixed size 100×40 pixels, half of which exactly showing a car and the other half showing other scenes or objects. Since the proposed localization approach has the benefit of requiring no heuristics about the possible locations and sizes of the bounding boxes, we use a test set consisting of 107 images with varying resolution containing 139 cars in sizes between 89×36 and 212×85. This dataset also includes ground truth annotation for the test images in the form of bounding rectangles for all the cars. The training set and the multi-scale test set are consistent with the setup used in [5].

A few sample test images of the dataset is shown in figure 1. Note that some test images contain multiple cars and partial occlusion may exist between different cars as well as between a car and a "noise" object, such as a bicyclist, a pedestrian or a tree.

Fig. 1 Sample images in the multi-scale car dataset

5.2 Metric

The localization performance is measured by recall and precision, the same way as in [1] and [5]. A hypothesized bounding box is counted as a correct detection if its location coordinates and size lie within an ellipsoid centered at the true coordinates and size. The axes of the ellipsoid are 25% of the true object dimensions in each direction. For multiple detected bounding boxes satisfying the above criteria for the same object, only one is counted as correct and the others are counted as false detections.

5.3 Gaussianized Vectors

The feature vectors for each image are extracted as follows. First, square patches randomly sized between 4×4 and 12×12 are extracted on a dense pixel grid. Second, an 128-dimensional SIFT vector is extracted from each of these square patches. Third, each SIFT vector is reduced to 64 dimensions by Principal Component Analysis. Therefore, each image is converted to a set of 64-dimensional feature vectors.

These feature vectors are further transformed into Gaussianized vector representations as described in Section 2. Each image is therefore represented as a Gaussianized vector. In particular, we carry out the experiment with 32, 64, 128 Gaussian components in the GMMs respectively.

5.4 Robustness to Within-Class Variation

We identify the subspace that contains the undesirable within-class variation using the eigen analysis method in Section 3. In particular, the subspace consists of top 100 dimensions, out of all the dimensions of the Gaussianized vectors, that are to be suppressed in the calculation of the kernels.

5.5 Results

To keep the setting the same as in [5], we search each test image for the three best bounding boxes, each affiliated with the quality function score. In particular, the branch-and-bound search scheme is applied to each test image three times. After each time, those features claimed by the found boxes are removed as discussed in Section 4.1.

The ROC curves, characterizing precision vs. recall, are obtained by changing the threshold on the quality function score for the found boxes. The equal error rate (EER) equals 1 − F-measure when precision equals recall.

The ROC curves and the EER are presented in Figure 2 and Figure 3 respectively. We compare the results with a localization system using the same banch-and-bound scheme, but based on the generic histogram of keywords with 1000 entry codebook generated from SURF descriptors at different scales on a dense pixel grid [5].

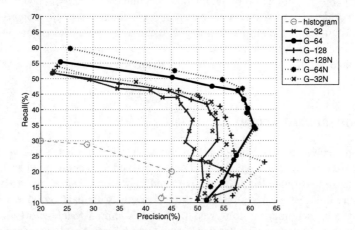

Fig. 2 ROC curves for multi-scale car detection. ("G-n" denotes the result using n components in the Gaussianized vector representation. The suffix "N" means the within-class normalization. "Histogram" denotes the performance using the generic histogram-of-keywords approach by Lampert et al.)

Efficient Object Localization with Variation-Normalized Gaussianized Vectors

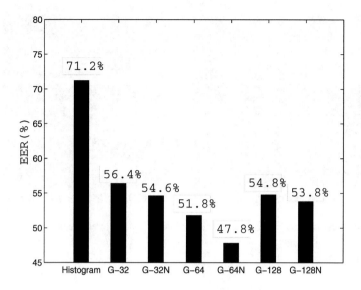

Fig. 3 Equal Error Rates for multi-scale car detection

We can see that the Gaussianized vector representation outperforms the histogram of keywords in this multi-scale object detection task. In particular, using 64 Gaussian components gives the best performance. In general, normalizing against within-class variation further improves the system.

In Figure 4, we present a few examples of correct detection and erroneous detection using 64 Gaussian components. Each test image is accompanied by a "per-feature-contribution" map. Negative and positive contributions are denoted by blue and red, with the color saturation reflecting absolute values. The quality function evaluated on a bounding box is the sum of all the per-feature-contributions, as discussed in Section 4.

The examples of correct detection demonstrate that the system can effectively localize one or multiple objects in complex backgrounds.

The three examples of erroneous detection probably occur for different reasons: 1) The car is a bit atypical, resulting in fewer features with highly positive contributions. 2) The two cars and some ground texture form one rectangle area with highly positive contributions, bigger than the two true bounding boxes. 3) The car is highly confusable with the background, resulting in too many highly negative contributions everywhere, preventing any rectangle to yield a high value for the quality function.

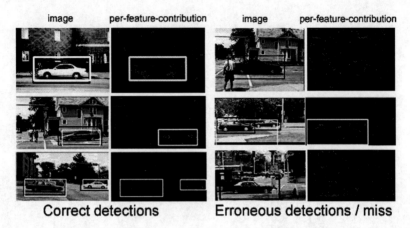

Fig. 4 Examples of good and bad localization based on Gaussianized vector representation. (The black and white bounding boxes in the images are the ground truth and the hypotheses respectively. Best viewed in color.)

6 Conclusion

In this chapter, we discuss effective object localization leveraging efficient and effective searching method, and robust image representation and learning method. In particular, we present an efficient object localization approach based on the Gaussianized vector representation. We design a quality bound for rectangle sets characterized by the Gaussianized vector representation. This bound can be obtained for any rectangle set within the image boundaries, with little extra computational cost, in addition to calculating the Gaussianized vector representation for the whole image classification problem. Adopting the branch-and-bound search scheme, we leverage the proposed quality bound for fast hierarchical search.

We further incorporate a normalization approach that suppresses the within-class variation, by de-emphasizing the undesirable subspace in the Gaussianized vector representation kernels. This helps achieve improved robustness to variation in the object class and the background.

The proposed object localization approach based on the Gaussianized vector representation outperforms a similar localization system based on the generic histogram-of-keywords representation on a multi-scale car dataset.

Acknowledgements. This research is funded by NSF grants IIS 08-03219 and IIS-0703624.

References

1. Agarwal, S., Awan, A., Roth, D.: Learning to detect objects in images via a sparse, part-based representation. IEEE Transactions on Pattern Analysis and Machine Intelligence 26(11), 1475–1490 (2004)
2. Dalal, N., Triggs, B.: Histograms of oriented gradients for human detection. In: CVPR, pp. 886–893 (2005)
3. Fei-Fei, L., Perona, P.: A Bayesian hierarchical model for learning natural scene categories. In: CVPR (2005)
4. Hatch, A., Stolcke, A.: Generalized linear kernels for one-versus-all classification: application to speaker recognition. In: ICASSP, vol. V, pp. 585–588 (2006)
5. Lampert, C., Blaschko, M., Hofmann, T.: Beyond sliding windows: Object localization by efficient subwindow search. In: Proc. of CVPR (2008)
6. Lazebnik, S., Schmid, C., Ponce, J.: Beyond bags of features: Spatial pyramid matching for recognizing natural scene categories. In: CVPR (2006)
7. Permuter, H., Francos, J., Jermyn, I.: Gaussian mixture models of texture and colour for image database retrieval. In: Proceedings of the IEEE International Conference on Acoustics, Speech, and Signal Processing (ICASSP 2003), vol. 3, pp. III-569-72 (April 2003)
8. Reynolds, D.A., Quatieri, T.F., Dunn, R.B.: Speaker verification using adapted gaussian mixture models. Digital Signal Processing 10, 19–41 (2000)
9. Rowley, H.A., Baluja, S., Kanade, T.: Human face detection in visual scenes. In: NIPS 8, pp. 875–881 (1996)
10. Viola, P., Jones, M.: Rapid object detection using a boosted cascade of simple features. In: Proc. of CVPR (2001)
11. Yan, S., Zhou, X., Liu, M., Hasegawa-Johnson, M., Huang, T.S.: Regression from patch-kernel. In: CVPR (2008)
12. Zhou, X., Zhuang, X., Tang, H., Hasegawa-Johnson, M., Huang, T.S.: A Novel Gaussianized Vector Representation for Natural Scene Categorization. In: ICPR (2008)
13. Zhou, X., Zhuang, X., Yan, S., Chang, S., Hasegawa-Johnson, M., Huang, T.S.: SIFT-Bag Kernel for Video Event Analysis. In: ACM Multimedia (2008)
14. Zhu, Q., Avidan, S., Yeh, M.-c., Cheng, K.-t.: Fast human detection using a cascade of histograms of oriented gradients. In: CVPR, pp. 1491–1498 (2006)
15. Zhuang, X., Zhou, X., Hasegawa-Johnson, M., Huang, T.S.: Face Age Estimation Using Patch-based Hidden Markov Model Supervectors. In: ICPR (2008)
16. Zhuang, X., Zhou, X., Hasegawa-Johnson, M., Huang, T.S.: Efficient object localization with gaussianized vector representation. In: *IMCE*: Proceedings of the 1st International Workshop on Interactive Multimedia for Consumer Electronics, pp. 89–96 (2009)

Fusion of Motion and Appearance for Robust People Detection in Cluttered Scenes

Jianguo Zhang and Shaogang Gong

Abstract. Robust detection of people in video is critical in visual surveillance. In this work we present a framework for robust people detection in highly cluttered scenes with low resolution image sequences. Our model utilises both human appearance and their long-term motion information through a fusion formulated in a Bayesian framework. In particular, we introduce a spatial pyramid Gaussian Mixture approach to model variations of long-term human motion information, which is computed via an improved background modeling using spatial motion constrains. Simultaneously, people appearance is modeled by histograms of oriented gradients. Experiments demonstrate that our method reduces significantly false positive rate compared to that of a state of the art human detector under very challenging lighting condition, occlusion and background clutter.

1 Introduction

Accurate and robust pedestrian detection in a busy public scene is an essential yet challenging task in visual surveillance. The difficulties lie in modelling both object and background clutter contributed by a host of factors including changing object appearance, diversity of pose and scale, moving background, occlusion, imaging noise, and lighting change. Usually pedestrians in public space are characterized by two dominant visual features: *appearance* and *motion*. There is a large body of

---

Jianguo Zhang
School of Electronics, Electrical Engineering and Computer Science,
Queen's University Belfast, BT7 1NN, UK
e-mail: jianguo.zhang@ieee.org

Shaogang Gong
School of Electronic Engineering and Computer Science,
Queen Mary University of London, London E1 4NS, UK
e-mail: sgg@dcs.qmul.ac.uk

work in human detection (see [7] and [2] for a survey). These work can be broadly categorized into two groups: static and dynamic people detectors. Static people detectors rely mainly on finding robust appearance features that allow human form to be discriminated against a cluttered background using a classifier such as SVM or AdaBoost searching through a set of sub-images by a sliding window. Typical features include rectified Haar wavelets [17], rectangular features [23], and SIFT (Scale Invariant Feature Transform) like features such as histogram of oriented gradients [16, 3]. Papageorgiou et al. [17] described a pedestrian detector based on SVM using Harr wavelet features. Gavrila and Philomin [6] presented a real-time pedestrian detection system by utilizing silhouettes information extracted from edge images. The candidate of the silhouettes is selected as the one with the smallest chamfer distance to a set of learned human shape examples. On the other hand, there is little progress on dynamic detectors, although the idea of using pure motion information for human pattern recognition is not new [11, 9, 20]. Most existing work utilises optic flow. Viola et al. [23] proposed a very efficient detector using AdaBoost that can achieve real-time performance. The rather simple rectangular features and the cascade structure account for the efficiency of this approach. Motion information was also taken into account through a coarse estimation of optic flow between two consecutive frames. Similar work of using optic flow for people detection can be found in [4]. To achieve satisfactory performance, this approach assumes that the human motion information in the test sequences is similar to those in the training set. Other related work using motion information includes human behavior recognition by distribution of 3D spatial-temporal interest points [22, 13], 3D volumetric features [12], or through 3D correlation [1]. Overall, existing methods for computing motion assume mostly that the motion is locally smooth. However this is untrue especially in busy public scenes when measuring optic flow is sensitive to noise and unreliable due to lighting change, reflection, moving background such as tree leaves (see Fig. 2).

To date, work on utilising both motion and appearance information remains in its infancy. To our best knowledge, there is little work performing direct people detection using both appearance and long-term motion information, whilst our previous work [25] has show some promising detection results using long-term motion score. In this work, we present a robust framework for people detection in highly cluttered public scenes by utilizing both human appearance and their long-term motion information whilst reliable optic flow cannot be estimated. We further introduce a spatial pyramid Gaussian Mixture approach to effectively model the variations of long-term motion information which takes into the account of local geometric constrains, and shows slightly better results than just using pure motion score [25]. Our method does not require the estimation of continuous motion such as optic flow in training thus reduces the number of features required for training a classifier. It allows for any detected appearance hypothesis to be verified using long-term motion history analysis. We show experimental results to demonstrate the efficiency and robustness of the proposed approach against that of a state of the art static people detector.

2 Methodology

In contrast to video sequences captured under well-controlled environment at frame rate, our task for people detection requires to work in highly cluttered public scene (underground) given low resolution data often at low frame rate. The scene also suffers from (1) significant lighting changes, which makes the motion estimation unstable and noisy; (2) heavy occlusions, which requires the people detector to handle partial match; (3) extensive background clutters, which can cause high false alarms. To this end, we propose a robust people detection method for video sequences by fusing static appearance feature based detector with a long-term motion based spatial pyramid likelihood measure. An overview of our method is shown in Fig. 1.

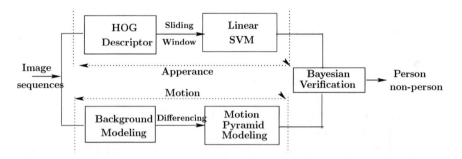

Fig. 1 Flow chart of our method for pedestrian detection. An appearance based detector is used to create the initial hypothesis and long-term motion is modeled by the motion pyramid approach. The above cues are combined in a Bayesian framework. The final candidates are selected by thresholding.

2.1 Generating Hypothesis

We adopt a static people detector proposed by Dalal and Triggs [3] to generate static human presence hypothesis in each frame. To achieve scale invariance, this detector utilizes a multi-scale sliding window approach, i.e. scanning each frame at each scale level. Each sub-window image patch centered at location i (denoted by v_i, where $i = 1 : n$ and n is the number of patches) is transformed into a feature vector before being classified into either human foreground or scene background by a classifier. The feature vector used here is a SIFT [16] like feature based on histogram of gradient orientation. The basic idea is that local object appearance and shape can often be characterized rather well by the distribution of local intensity gradients or edge directions, even without any precise knowledge of corresponding gradient or edge positions (similar work can be found in [21] using histograms of scale normalized, oriented derivatives to detect and recognize arbitrary object classes). The size of the detection window is 32×64 including 8 pixels of margin beyond the window size. A linear SVM is used as the classifier and the output of the classifier serves as the confidence measure for our static human appearance hypothesis. This approach has achieved very good detection rate in static images of outdoor scene [3], e.g. image samples from the MIT pedestrian dataset [17]. However, the lack of motion information makes this detector less robust to background clutters. This problem becomes severe in cluttered scenes with poor lighting, such as in public underground, when such a static human detector gives unacceptable false alarm rate, as shown by examples in Figs 8 and 9. Simply increasing the threshold of the score for generating the hypothesis does not result in reducing the false alarms because in such cluttered scenes, regions in background have very similar appearance to that of people, e.g. as shown in Fig. 8 (e). Here the false alarms on the wall have very high scores produced by the classifier and do indeed look like standing people. Similar observations can be found in the example shown in Fig. 9 (b). Given that in any public scene, people exhibit inevitably long-term-moving patterns instead of just a static pattern, we consider a detection model based on fusing detected static human presence hypothesis with their long-term motion history information as follows.

2.2 Motion Confidence

One way to utilize motion information is to compute optic flow [10, 18, 5]. However, optic flow estimation makes a strong assumption that motions are only caused by either relative movement between the camera and the object of interest or ego-motion. The accuracy of flow estimation is based on well sampled data, i.e. local smoothness. However, both assumptions are not usually satisfied in real applications. First, large lighting changes usually result in noisy flow field. Second, relatively fast action w.r.t the camera, i.e. low frame rate, also results in highly discontinuous motion which is far from smooth. Examples of estimating optic flow in a underground scene are shown in Fig. 2. The optic flow was computed using a robust method proposed by Gautama et al. [5]. However, it is evident that the resulting flow field is very

Fig. 2 Optic flow estimation comparison in different sequences. (a) regular flow on well-captured outdoor sequence; (b) noisy flow on a real underground sequence. Note that in (b), the upper-right corner has some distinct noisy optic flow caused by lighting changes and object reflections.

noisy and unstable. To address this problem, in this work we adopt an alternative long-term motion estimation approach using background extraction and subtraction, given that most surveillance CCVT systems are based on fixed views. More precisely, we utilise a Gaussian mixture background model of [24]:

$$b(x,y) = \sum_i \alpha_i g(f(x,y), \theta_{i,x,y}, \sigma_{i,x,y}), \qquad (1)$$

where x, y is the location of each pixel, $(\theta_{i,x,y}, \sigma_{i,x,y})$ are the model parameters of each individual Gaussian component g, and $f_t(x,y)$ is the local pixel intensity. Once the parameters are estimated, the likelihood of one frame $f(x,y)$ at time t with respect to the background model is computed as the probability distance given by

$$v(x,y) = \sum_i \alpha_i \exp\left(-\frac{1}{2} \frac{(f_t(x,y) - \theta_{i,x,y})^2}{\sigma_{i,x,y}^2}\right) \qquad (2)$$

This type of motion information is very effective at highlighting changes in motion of every pixel in the scene. However, this is also an undesirable property since the noisy motion caused by lighting changes is inevitably augmented. See Fig. 8 (b) as an example. To suppress the noisy motion caused by lighting changes, we further take spatial motion contrast into consideration in the Gaussian mixture model as follows:

$$v(x,y) = \sum_i \alpha_i \exp\left(-\frac{1}{2} \frac{(f_t(x,y) - \theta_{i,x,y})^2}{\sigma_s^2}\right) \qquad (3)$$

In the background model of Eq.(2), $\sigma_{i,x,y}$ is the estimated strength of the motion of each pixel at (x,y), we calculate σ_s in Eq.(3) as the mean of $\sigma_{i,x,y}$. Examples of motion extraction using this model are shown in Fig. 8 (b) and (c), where in (b) motion was estimated using the Gaussian mixture background model without considering spatial motion contrast whilst in (c), it was taken into account. This demonstrates clearly the effectiveness of utilising the spatial motion contrast measure given by Eq.(3) for removing motion noise as compared to existing Gaussian mixture models.

2.3 Spatial Motion Descriptor

Base on the background modeling described in Sec. 2.2, we can estimate a motion confidence measure of each hypothesis created by the static detector. Next is to construct a robust hypothesis descriptor based on the motion information. Inspired by the success of SIFT descriptor [16], we propose a multi-level spatial pyramid descriptor by directly utilising the motion confidence calculated from Eq.(3) to effectively describe the motion region of the hypothesis. The descriptor extraction procedure consists of the following steps:

1. Creating a codebook of confidence measure. Because the confidence $v(x,y)$ is in principle a probability with $v \in [0,1]$, we can create C bins of the value with

equal width. Each bin center is treated as a code, denoted as $w_c, c = 1, 2, 3, ..., C$. In this paper, we experimentally set the value of C as 20.

2. Building a soft codebook histogram where each pixel contributes to more than one bins. The contribution received by bin c is calculated according to Eq.(4).

$$h(c) = \sum_i exp(\frac{(v_i - w_c)^2}{\sigma}) \qquad (4)$$

Where σ is set to the half of the bin width. i is the index of points with the region of interest for a given hypothesis. The histogram is then normalized to unit $L1$ norm.

3. Building a spatial pyramid at different levels by partitioning the hypothesis region into different sub-regions as shown in Fig. 3. For each sub-region, we build a codebook histogram. Then we concatenate the histogram of all the sub-regions at the same level into a feature vector. Thus a motion descriptor is represented as a set of feature vectors across different pyramid levels, e.g. for a pyramid with 3 levels, the motion descriptor is $\{Y(l = 0), Y(l = 1), Y(l = 2)\}$ (l is the pyramid level) with dimensionality at each level being $\{20, 80, 320\}$ when the length of the codebook is 20.

There are several merits of the proposed descriptor. First it uses directly a confidence measure from the background modeling, thus avoids the computation of additional features and reduces the computational cost. Secondly, it takes into account the geometrical interactions between the local parts of a human body configuration, thus makes it more effective.

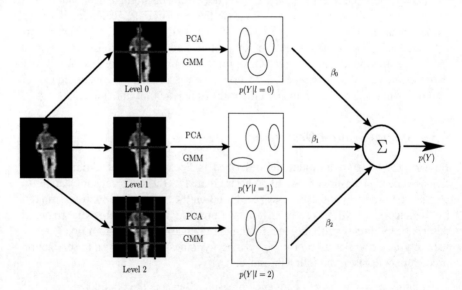

Fig. 3 Illustration of the modeling process of the pyramid Gaussian Mixture with 3 levels

To improve robustness of the descriptor against error in the location of a hypothesis, the motion confidence within each hypothesis is first Gaussian weighted. In this way, we give higher weight to the contribution of motion near the center of the hypothesis window, whilst less emphasis is placed to further away from the center. The Gaussian weighting mask we used in our experiments is as follows. We set it to the window size of a hypothesis created by the static human detector, where $g(x,y)$ is an anisotropic Gaussian envelope given as:

$$g(x,y) = \frac{1}{Z}exp\left(\frac{1}{2}([x,y]-[x_0,y_0])\Sigma^{-1}([x,y]-[x_0,y_0])^T\right) \quad (5)$$

where $[x_0, y_0]$ is the center point of the region, $[x, y]$ is the spatial coordinates of the points within the local region, and Z is the normalization factor. Σ is the spatial correlation matrix. We set as $\begin{bmatrix} 1/2w & 0 \\ 0 & 1/2h \end{bmatrix}$, where w, h are the width and height of the hypothesis window. Examples of measuring the Gaussian weighting mask against the corresponding bounding boxes are shown in Fig. 10. To further make the descriptor invariant to scale changes, we normalize the hypothesis regions into a region with of the same size of 32x32 before partitioning the motion template.

2.4 Pyramid Bayesian Verification

Pyramid GMM Modeling. In order to cope with variation in motion under different viewpoints and human poses, we need to model the long-term motion distribution of people. To this end, we collect a set of motion templates of human after background modeling as in Eq. (3), denoted by $q_i, i = 1, 2, ..., n$. The corresponding motion descriptors are represented by $Y_i, i = 1, 2, ..., n$. On the other hand, we want to incorporate the spatial constrains at different levels in the pyramid, we introduce a *spatial pyramid Gaussian mixture modeling (GMM) approach*. The idea is that for each pyramid level, we model the conditional distribution of descriptors as a Gaussian Mixture, denoted as $p(Y|l)$, and the whole pyramid GMM model is expressed as the combination of those GMMs with different weights. Mathematically, the model of descriptor Y can be written probabilistically as follows:

$$p(Y) = \Sigma_l p(Y|l)p(l) \quad (6)$$

In this model, l denotes the pyramid level and $p(l)$ is the prior associated with each level (the setting of its value will be explained later in this section). The conditional distribution $p(Y|l)$ w.r.t. l is specifically modeled as a GMM:

$$p(Y|l) = \sum_{i \in I(l)} \alpha_i N(y(l); u_i, \Sigma_i) \quad (7)$$

Where $I(l)$ is the number of components in a GMM at level l and $y(l)$ represents the motion descriptor at level l, i.e., $Y(l = i, i \in 0, 1, ..., L)$. $(\alpha_i, u_i, \Sigma_i)$ are the parameters of Gaussian mixtures.

Item $P(l)$ in Eq. (6) is the prior associated with each pyramid level and used to combine the the conditional likelihood for each level, denoted as β_l. Motivated by the theory of spatial pyramid matching, we determine the weight β_l from the formulation of a maximum-weight problem [8, 14], which is inversely proportional to cell width at that level as shown in Eq. (8). Intuitively, we want to penalize likelihood found in larger cells because they involve increasingly dissimilar features. Taken all of these into consideration, we calculate a pyramid likelihood as follows:

$$p(Y) = \sum_{l=0}^{L} \beta_l p(Y|l) = \frac{1}{2^L} p(Y|l=0) + \sum_{l=1}^{L} \left(\frac{1}{2^{L-l+1}} p_l(Y|l) \right) \quad (8)$$

It is worthwhile noting that our motion descriptor has a high dimensionality, e.g. the dimensions of the descriptor at level 2 is 320. Thus directly using Gaussian Mixture Model would involve the estimation of thousands of parameters. Typically, this would be time consuming and quite unstable in a higher dimensional space. To avoid this, we first use Principle Component Analysis (PCA) to reduce the dimensionality of the motion descriptor for each pyramid level respectively, and then use GMM to model its distributions. The effectiveness of PCA-GMM has been demonstrated in [15]. In our work here, GMM is learned by maximizing the likelihood function using Expectation-Maximization. The number of the Gaussian components is determined automatically using Minimum Description Length criterion [19]. Thus the modeling process of the pyramid Gaussian Mixture is illustrated in Fig. 3. Once the parameters ($\hat{\Theta}$) are estimated, given a motion template m with pyramid motion descriptor Y_t under a people hypothesis h, the likelihood with respect to the learned human model is calculated as follows:

$$\begin{aligned} p(m|h,o) &= p(Y_t|\hat{\Theta}) \\ &= \sum_l \beta_l p(Y_t|l, \hat{\theta}_i) \\ &= \sum_l \beta_l \sum_{i \in I(l)} \hat{\alpha}_i N(Y_t(l); \hat{u}_i, \hat{\Sigma}_i) \end{aligned} \quad (9)$$

where $(\hat{\alpha}_i, \hat{u}_i, \hat{\Sigma}_i)$ are the estimated parameters of Gaussian mixture. Since Y_t is computed based on the a given motion template (m) of a hypothesis (h) by a static object class detector (o), it is natural to represent $p(Y_t|\hat{\Theta})$ as $p(m|h,o)$. In the following section, we will deploy these annotations to give a clear illustration of the verification process using the Bayesian graph model.

Verification. For a bounding box hypothesis, we wish to find the probability of the presence of an object given its motion template m and appearance measure c, $p(o|c,m,h)$, which is given by the Bayesian rule as follows:

$$p(o|c,m,h) = \frac{1}{Z} p(m|h,o) p(c|h,o) p(h|o) \quad (10)$$

where Z is the normalization factor. In this model, we assume that the motion m and the appearance c are conditionally independent. To understand this more clearly, the directed probability graph model of the Bayesian verification process (Eq. (10)) is shown in Fig. 4, where the arrows indicate the dependencies between variables. $p(m|h,o)$ is the contribution of the motion within the hypothesis bounding box given

Fig. 4 The graph models of the Bayesian verification process

the object, which is computed using Eq.(9). $p(c|h,o)$ is the appearance confidence measure generated by a static people detector (here, we use logistic regression to convert the output of the detector into a probabilistic interpretation). $p(h|o)$ is the confidence of the hypothesis by the object detector, which in our case is a hypothesis presence indicator for a certain object class. The final candidates are selected by thresholding $p(o|c,m,h)$.

3 Experimental Results

3.1 Data Set

Training set: We use INRIA image dataset[1] for training, which is totally independent to our test video data. The INRIA dataset is both challenging and also does not contain any motion information. It contains 607 positive training images, together with their left-right reflections (1214 images in all). A fixed set of 12180 patches are randomly sampled from 1218 person-free training images. Examples of these images are shown in Fig. 5. 300 human motion templates are collected from a separate courtyard sequences, which are used to learn the parameters of the spatial pyramid Gaussian Mixture. Examples of those are illustrated in Fig. 7.

Test set: Our test set contains image frames from CCTV video sequences taken from underground stations and platforms by fixed cameras. One set of images is from a train platform containing 3710 frames. The other set is from a ticket office area containing 160 frames. In contrast to other video sequences captured under well-controlled environment, these sequences present significant lighting changes and background clutter. Many frames contain multiple people under severe occlusions. Fig. 6 shows 6 consecutive example frames from the platform scene.

3.2 Detection Results

Examples of detections at the platform area scene from our dynamic detector are compared and shown against those from a static people detector [3] in Fig. 8. The detected boxes are displayed on each frame. Fig. 8 (d) shows detection results from

---

[1] The dataset may be downloaded from http://lear.inrialpes.fr/data.

Fig. 5 Positive and negative training examples used in our experiments

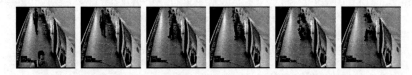

Fig. 6 Six consecutive frames from the test sequences

Fig. 7 Examples of motion templates used in our experiments to learn the spatial pyramid Gaussian mixture model.

the static detector whilst Fig. 8 (e) and (f) show the detections of our dynamic detector. Fig. 8 (f) is an improved version of Fig. 8 (e) by increasing the robustness of motion estimation using spatial contrast measure given in Eq.(3). Thus it further removes additional false alarms. Another detection examples from the sequences of the underground ticket office are shown in Fig. 9. The dataset presents a lot of background clutters where ticket machines appear to be similar to the appearance of people standing there. So it is reasonable for the static detectors to produce false detections at where some ticket machines are located, as shown in Fig. 9 (b). Our dynamic detector has shown to be able to remove those false alarms.

To quantify the detection performance, we also perform a quantitative evaluation of both the dynamic detector and static detector on the ticket-office scene from

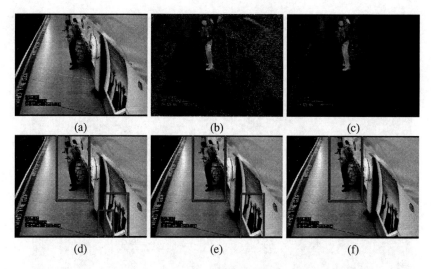

Fig. 8 Examples of human detection in each step. (a) the original slice; (b) initial motion confidence map only using Gaussian mixture; (c) refined motion confidence map; (d) initial hypothesises by using pure static human detector; (e) detection results by using the motion map of (b); (f) refined detection results by using the motion map of (c).

which manually labelled ground truth was available. By varying the threshold of the detection scores one at a time, we obtain the Receiver Operating Characteristics (ROC) curves of those detectors, showing false positive rate versus true positive rate (see Fig 11). When comparing with the ground truth annotation, we measured the overlap score between the detected bounding box and ground true bounding box. A detection with overlap score larger than 50% is labeled as a 'match'. We add the ROC curve of the approach using motion score directly in the Bayesian verification process as done in [25]. For further explaining the role of motion information played in improving the detection rate, we also plotted the ROC curve of a pure motion detector, i.e. the bounding boxes are weighted only by their motion information as computed by Eq.(3). From this experiment, it is clear that the motion information plays a critical role in accurate detection. Simply using the motion information alone also gave good detection as shown by the ROC curve. This is because most of the motions in this particular scene were caused actually by human movement. The ROC curve shows that our dynamic detector improves significantly the performance of the static detector by Dalal and Triggs [3] and using the proposed pyramid Gaussian Mixture modeling performs better than just using the motion score. The false alarms rate has been greatly reduced compared to the static detector. For example, to achieve a detection rate of 70% on the ticket-office scene, our detector produces 110 false alarms whilst the detector by Dalal and Triggs generated 370 false alarms, over 3 times more.

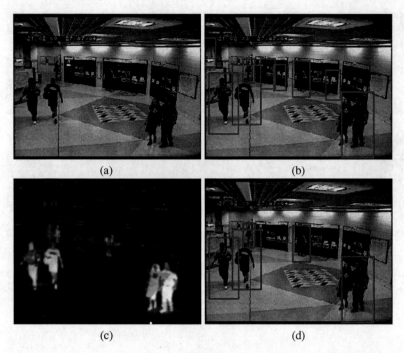

Fig. 9 Examples of human detection in the ticket office with heavy background clutter. (a) the original slice; (b) initial hypothesises use pure static human detector; (c) motion confidence map; (d) detection results using the motion map of (c).

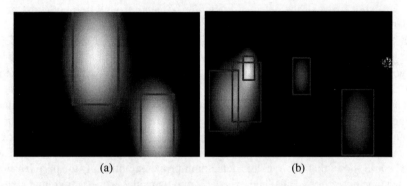

Fig. 10 Gaussian weighting masks corresponding to the hypothesises bounding box. (a) corresponds to Fig. 8(e); (b) corresponds to Fig. 9(d).

4 Discussion and Conclusions

In this paper, we presented a framework for robust people detection in highly cluttered scenes with low resolution image sequences. In particular, we introduced a

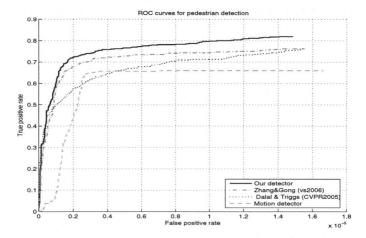

Fig. 11 ROC curves on the sequences of the underground ticket office for both the dynamic and static people detectors as well as a pure motion based people detector.

spatial pyramid Gaussian Mixture approach to model variations in long-term motion of human movement, which is computed via an improved background modeling. Our model utilises both human appearance and their long-term motion information through a fusion formulated in a Bayesian framework. Experiments demonstrate that our method reduces significantly the false positive rate compared to that of the state of the art static human detector under very challenging conditions. Note that our current people appearance is modeled by histograms of oriented gradients, which is an extension of the static detector proposed by Dalal and Triggs. However, in principle, any other type of static detectors can be used in our framework. As our model is based on long-term motion information therefore it requires a fixed camera view during detection. Building a hybrid model of both long-term and short-term motion information could possibly give more robust detections and also be adaptive to some background and viewpoint change.

References

1. Boiman, O., Irani, M.: Detecting irregularities in images and in video. In: International Conference on Computer Vision, pp. 462–469 (2005)
2. Cutler, R., Davis, L.: Robust real-time periodic motion detection: Analysis and applications. IEEE Transactions on Pattern Analysis and Machine Intelligence 22(8), 781–796 (2000)
3. Dalal, N., Triggs, B.: Histograms of oriented gradients for human detection. In: International Conference on Computer Vision & Pattern Recognition, vol. 2, pp. 886–893 (2005)
4. Dalal, N., Triggs, B., Schmid, C.: Human detection using oriented histograms of flow and appearance. In: Leonardis, A., Bischof, H., Pinz, A. (eds.) ECCV 2006. LNCS, vol. 3952, pp. 428–441. Springer, Heidelberg (2006)

5. Gautama, T., Van Hulle, M.: A phase-based approach to the estimation of the optical flow field using spatial filtering. IEEE Transactions on Neural Networks 13, 1127–1136 (2002)
6. Gavrila, D., Philomin, V.: Real-time object detection for "smart" vehicles. In: IEEE Conference on Computer Vision and Pattern Recognition, pp. 87–93 (1999)
7. Gavrila, D.M.: The visual analysis of human movement: A survey. Computer Vision and Image Understanding 73(1), 82–98 (1999)
8. Grauman, K., Darrell, T.: Pyramid match kernel: Discriminative classification with sets of image features. In: International Conference on Computer Vision, pp. 1458–1465 (2005)
9. Hoffman, D.D., Flinchbaugh, B.E.: The interpretation of biological motion. Biological Cybernetics 42, 195–204 (1982)
10. Horn, B.K.P., Schunck, B.G.: "determining optical flow": A retrospective. Artifical Intelligence 59(1-2), 81–87 (1993)
11. Johansson, G.: Visual perception of biological motion and a model for its analysis. Perception and Psychophysics 14, 201–211 (1973)
12. Ke, Y., Sukthankar, R., Hebert, M.: Efficient visual event detection using volumetric features. In: International Conference on Computer Vision, pp. 166–173 (2005)
13. Laptev, I.: On space-time interest points. International Journal of Computer Vision 64(2), 107–123 (2005)
14. Lazebnik, S., Schmid, C., Ponce, J.: Beyond bags of features: Spatial pyramid matching for recognizing natural scene categories. In: IEEE Conference on Computer Vision and Pattern Recognition, pp. 2169–2178 (2006)
15. de Lima, C., Alcaim, A., Apolinario, J.J.: On the use of pca in gmm and ar-vector models for text independent speaker verification. In: International Conference on Digital Signal Processing, vol. 2 (2002)
16. Lowe, D.G.: Distinctive image features from scale-invariant keypoints. International Journal of Computer Vision 60(2), 91–110 (2004)
17. Papageorgiou, C., Poggio, T.: A trainable system for object detection. International Journal of Computer Vision 38(1), 15–33 (2000)
18. Proesmans, M., Gool, L.J.V., Pauwels, E.J., Oosterlinck, A.: Determination of optical flow and its discontinuities using non-linear diffusion. In: Eklundh, J.-O. (ed.) ECCV 1994. LNCS, vol. 801, pp. 295–304. Springer, Heidelberg (1994)
19. Rissanen, J.: A universal prior for integers and estimation by minimum description length. The Annals of Statistics, 416–431 (1983)
20. Sankaranarayanan, A., Chellappa, R., Zheng, Q.: Tracking objects in video using motion and appearance models. In: IEEE International Conference on Image Processing, vol. 2, pp. 394–397 (2005)
21. Schiele, B., Crowley, J.: Recognition without correspondence using multidimensional receptive field histograms. International Journal of Computer Vision 36(1), 31–50 (2000)
22. Schuldt, C., Laptev, I., Caputo, B.: Recognizing human actions: A local svm approach. In: International Conference on Pattern Recognition, Cambridge, UK, pp. 32–36 (2004)
23. Viola, P., Jones, M.J., Snow, D.: Detecting pedestrians using patterns of motion and appearance. International Journal of Computer Vision 63(2), 153–161 (2005)
24. Xiang, T., Gong, S.: Beyond tracking: Modelling activity and understanding behaviour. International Journal of Computer Vision 67(1), 21–51 (2006)
25. Zhang, J., Gong, S.: Beyond static detectors: A bayesian approach to fusing long-term motion with appearance for robust people detection in highly cluttered scenes. In: IEEE Workshop on Visual Surveillance in conjunction with ECCV 2006, Graz, pp. 121–128 (2006)

Understanding Sports Video Using Players Trajectories

Alexandre Hervieu and Patrick Bouthemy

Abstract. One of the main goal for novel machine learning and computer vision systems is to perform automatic video event understanding. In this chapter, we present a content-based approach for understanding sports videos using players trajectories. To this aim, an object-based approach for temporal analysis of videos is described. An original hierarchical parallel semi-Markov model (HPaSMM) is proposed. In this latter, a lower level is used to model players trajectories motions and interactions using parallel hidden Markov models, while an upper level relying on semi-Markov chains is considered to describe activity phases. Such probabilistic graphical models help taking into account low level temporal causalities of trajectories features as well as upper level temporal transitions between activity phases. Hence, it provides an efficient and extensible machine learning tool for applications of sports video semantic-based understanding such that segmentation, summarization and indexing. To illustrate the efficiency of the proposed modeling, application of the novel modeling to two sports, and the corresponding results, are reported.

1 Introduction

Semantic-based interpretation of videos mainly consists in recognizing objects behaviors, where a behavior can be defined as a sequence of activities. Hence, the issue

Alexandre Hervieu
Fundació Barcelona Media Universitat Pompeu Fabra, Avenida Diagonal, 177
08017 Barcelona, Spain
e-mail: ahervieu@gmail.com

Patrick Bouthemy
INRIA, Centre Rennes - Bretagne Atlantique, Campus Universitaire de Beaulieu
35042 Rennes, France
e-mail: Patrick.Bouthemy@irisa.fr

here is to be able to detect activities as well as transitions between activities. Such an understanding is now of great interest in the computer vision field. It is motivated by several applications in various domains: video surveillance, sports video indexing and exploitation, video on demand... Trajectories of mobile video objects are now available with tracking systems and are prone to be exploited for video understanding. Indeed, trajectories provide a high level description of dynamical contents observed in videos.

Several approaches have been proposed to exploit mobile object trajectories for content-based video analysis [23, 1, 12]. Günsel et al. developed a video indexing framework based on the analysis of "video objects" [7]. Their method relies both on the dynamics of the tracked objects and on the interactions between the corresponding trajectories. Based on such interactions, other algorithms were developed to handle video content understanding. Let us mention a system that models interactions between moving entities in a video surveillance context, relying on Coupled Hidden Markov Models, that was proposed by Oliver et al. [19]. Other contributions describe trajectory-based schemes for activity recognition using the definition of scenarios within multi-agent Semi-Markov Chains (SMCs) [11, 17, 18].

In practice, however, due to the extreme variety of observable video content, no practical extensive application of these methods has yet been reported. That's why, in the sequel, we will concentrate on the analysis of video trajectories in the particular context of sports videos. Indeed, such activities are supervised by a set of rules while occurring in *a priori* known (most likely closed) spaces, providing hence a suitable experimentation field for semantic video understanding methods [13]. Moreover, since most sports (tennis, soccer, rugby, handball...) take place on a 2D ground plane, players movements are heavily characterized by their 2D trajectories in the court plane. Such trajectories, that are now extractable using computer vision tracking tools from the existing and abundant literature, hence provide rich semantic information on the observed sports videos.

More precisely thus, several approaches relying on video trajectories have been provided for sports video activity recognition [9]. However, such a method only provide classification of activities in video shots into two categories: "normal" and "unexpected" events. One trajectory-based framework, still, produces a semantic segmentation of sports video into activities. It was proposed by Perše et al. and is dedicated to basketball video analysis method [22]. Here, a first process segments tracked players trajectories into three different classes of basketball activities (offense, defense and time-out), relying on Gaussian mixtures and an *EM* algorithm trained on manually labeled sequences. Then, based on a partition of the court, a second stage achieves a template-based activity recognition of the offense video segments into three different classes of basketball play: *screen*, *move* and *player formation*. If not straightforward, extending such a method to other sports seems conceivable but would most certainly require some further investigations. On the contrary, in the following of this chapter, our aim is to provide an already general trajectory-based framework which can easily be extended and applied to most team and racquet sports.

To perform motion video interpretation using such trajectories, a general and original hierarchical parallel semi-Markov model (HPaSMM) is here proposed. A lower level is used to model players motions and interactions using parallel hidden Markov models, while an upper level relying on semi-Markov chains is considered to describe activity phases. This probabilistic framework helps handling temporal causalities of low level features within an upper level architecture that models temporal transitions between activity phases. Hence, such a modeling provides an efficient and extensible machine learning tool for sports semantic-based video event applications such that segmentation, summarization or indexing.

Spatio-temporal features modeled in the lower layer and used to model visual semantics (*i.e.*, the observed activity phases) are estimated from trajectory data. In order to proceed activity recognition under motion clutters, a kernel-based filtering is proposed to handle noisy spatio-temporal features. The considered features characterize on one side single players motions using velocity information, while distances between trajectories are used to describe interactions between players.

To show the versatility and efficiency of the proposed modeling, the methodology described in the first part of this paper is then applied to two very different sports: handball and squash. In order to represent squash activities, three features and two activity phases have been defined on the players trajectories, whereas handball video interpretation relies on five different trajectory-based features and eight activity phases. In the video context however and in order to propose a system that may be able to process a large variety of sports video content, invariance of the activity feature representation to some appropriate transformations is considered. Evaluations of our sports event reasoning method has been performed on real video sequences, and satisfying activity understanding results have been reached on both sports videos. We also have favorably compared with the hierarchical parallel hidden Markov models method (HPaHMM). In our previous works on sports video understanding [10], such a method was already considered for handball. A quite similar scheme was also considered, with a different low level feature modeling though, for squash video processing [8] . In this chapter however, we propose a generalization of the method and describes an overall algorithm for sports video understanding that may easily be extended to other team and racquet sports. We also go into important details in depth while providing updated results, comparisons and perspectives.

The structure of the chapter stands as follows. In Section 2, we will describe the hierarchical parallel semi-Markovian framework proposed for trajectory-based sport video understanding. Application of this framework to squash video processing is presented in Section 3 as well as corresponding experiments and results. Section 4 further describes the use of the novel modeling to a more complex set of activities for handball video understanding. Experimental results are also reported and discussed. In conclusion of this work, perspectives of extension to "general" high level analysis of videos are considered.

2 A Hierarchical Parallel Semi-Markovian Framework (HPaSMM) for Trajectory-Based Sport Video Understanding

To model activity phases observed in sports videos, we propose a hierarchical parallel semi-Markov model (HPaSMM). The proposed modeling is based upon a two-level hierarchy. Upper level layer is modeled using a semi-Markovian model [6]. Each upper level state of the HPaSMM correspond to an activity phase. In the lower layer, feature vectors are modeled using parallel hidden Markov models (PaHMMs, see [25]). These PaHMMs are used to characterize the upper level states of the HPaSMM.

Figure 1 contains an example of proposed framework which will be used in the following of this section to describe HPaSMMs. Upper level states are here denoted by S_i.

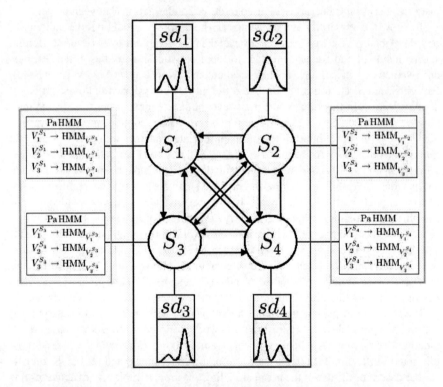

Fig. 1 Example of a HPaSMM architecture composed of $N = 4$ upper level states S_i. Each state corresponds to a given phase of activity. A phase of activity is modeled by a n-layer PaHMM (one layer for each feature vector, here $n = 3$) and by a GMM modeling state durations (sd_i is the state duration associated to state S_i). Upper level layer is surrounded by a red square and lower level layer is surrounded by green squares.

2.1 Upper Layer: Semi-Markovian Activity Modeling

Hence, in our HPaSMM framework, the upper level layer is dedicated to the modeling of activities. Each HPaSMM states S_i actually corresponds to an activity phase. This upper level layer is constructed using a semi-Markovian scheme [6] which has the advantage to explicitly model state durations in the upper level states S_i. Indeed, Markovian models stand upon the hypothesis that state durations follow geometrical laws. Yet, upper level states (*i.e.*, activities) do not necessarily follow simple geometrical laws and require more complex modelings. So, in the proposed HPaSMM, mixtures of Gaussian models (GMMs) are used to model the durations of the activity phases denoted by sd_i. In the illustration in Figure 1, the upper level layer is surrounded in red and is composed of four upper level states S_1, S_2, S_3, and S_4.

The set of parameters involved in the upper level layer is composed of A and ψ, where $A = \{a_{i,j}\}$ is the upper level HPaSMM state transition matrix at the time index of activity changes (see [6]), and ψ denotes the set of parameter related to GMMs state duration models.

2.2 Lower Layer: Parallel Markovian Feature Modeling

Lower level layer of the HPaSMM scheme is dedicated to the modeling of feature values computed on the trajectories. For a given sport, a number n of feature value v_j is chosen. They have to describe players movements and their respective interactions in each video frame. Moreover, to process a large variety of sports videos, such feature values will also have to stand invariant to appropriate transformations. Description of squash and handball feature representations are further provided in Sections 3 and 4.

The set of successive feature values v_j (one for each video frame) corresponding to a upper level state S_i are placed into a vector denoted by $V_j^{S_i}$. For each state of the upper level S_i, the corresponding feature vectors $V_j^{S_i}$ are modeled using a n-layer PaHMMs, where one HMM is dedicated to each feature vector $V_j^{S_i}$. The PaHMMs are here defined in a similar way than those presented in [25], where conditional probabilities B of observations are fitted by GMMs. In the illustration in Figure 1, the lower level layer is surrounded in green. For each considered activity S_i, three feature vectors $V_j^{S_i}$ are considered and modeled by a 3-layer PaHMM.

The set of PaHMMs parameters is denoted by ϕ. ϕ is composed of B, A' (state transition matrix) and π (initial state distribution) of the n-layer PaHMMs for each activity phase S_i.

Kernel approximation

We describe here a pre-processing that may be computed on the raw trajectory data to obtain continuous representations of the trajectories. Two advantages are implied in the choice of such a pre-processing. First of all, if trajectory data obtained from tracking procedure is noise corrupted, the kernel approximation help handling such issues by smoothing the observed trajectories. Moreover, as we will see in Section 3, low level feature values used to describe trajectories may be composed of velocity values that require computation of differential values. Using a kernel approximation of the trajectories provides a simple way to get such differential values.

Let us consider a trajectory T_k composed of a set of l_k points corresponding to the temporal successive positions of the moving object in the image plane, *i.e.*, $T_k = \{(x_{1,k}, y_{1,k}), ..., (x_{l_k,k}, y_{l_k,k})\}$. A continuous approximation of the trajectory T_k is defined by $\{(u_{t,k}, v_{t,k})\}_{t \in [1;l_k]}$ where:

$$u_{t,k} = \frac{\sum_{j=1}^{l_k} e^{-(\frac{t-j}{h})^2} x_{j,k}}{\sum_{j=1}^{l_k} e^{-(\frac{t-j}{h})^2}}, \ v_{t,k} = \frac{\sum_{j=1}^{l_k} e^{-(\frac{t-j}{h})^2} y_{j,k}}{\sum_{j=1}^{l_k} e^{-(\frac{t-j}{h})^2}}.$$

First and second order differential values $\dot{u}_{t,k}, \dot{v}_{t,k}, \ddot{u}_{t,k}$ and $\ddot{v}_{t,k}$ are easily obtained using standard derivation formulas. Yet simple, such calculations leads to long analytic expressions that are not detailed here.

Trajectory data used for experiments in Sections 3 and 4 are quite precise and does not need a denoising pre-processing, However, such a pre-processing is considered in Section 3 since differential values are needed for squash video processing. On the contrary, feature representation for handball video processing (see Section 4) does not require such a procedure.

Computation time reduction

To reduce computation time, a grouping procedure of the feature values may be performed. For the n considered feature vectors $V_j^{S_i}$, groups of k_{group} consecutive values are formed. For each of these groups, the mean values are computed and used to construct n new feature vectors $V_j^{S_i}$ (having sizes k_{group} times smaller than the original feature vectors).

2.3 Model Parameter Estimation

The entire set of parameter associated with a HPaSMM is finally given by $\theta = \{A, \phi, \psi\}$ and is estimated by a supervised learning stage. Training videos are considered and used to compute any parameter of θ. θ is further used to perform temporal activity phases recognition using a Viterbi algorithm (see Section 2.4).

A is the upper level HPaSMM state transition matrix. It is computed using observed transitions between activity phases in the training videos.

In the proposed HPaSMM, mixtures of Gaussian models (GMMs) are used to model durations of the activity phases S_i denoted by sd_i. The corresponding set of parameter ψ is obtained by fitting GMMs using "forward-backward" procedures. Initializations are here performed using a classical k-means algorithm.

PaHMMs sets of parameters are obtained using "Expectation-Maximization" procedures as it is described in [25].

2.4 Activity Recognition by Log-Likelihood Maximization

We give here details of the modified Viterbi algorithm when applied to log-likelihood maximization of HPaSMM. We consider here an upper level state sequence S containing R successive segments. Each segment corresponds to an activity associated with an upper level state S_i. We denote by q_r the time index of the end-point of the rth segment. The sequence of observations that characterize a segment r is given by $y_{(q_{r-1}+1, q_r]} = y_{q_{r-1}+1}, \ldots, y_{q_r}$ such that $S_{q_{r-1}+1} = \ldots = S_{q_r}$. θ is supposed known and is used to perform recognition of successive activity phases using the modified Viterbi algorithm as follows.

The algorithm provides the decoded sequence of upper level HPaSMM states $\hat{S}$ that maximizes the log-likelihood, *i.e.*, such that $\hat{S} = \arg\max_S \log P(y, S|\theta)$. The likelihood $P(y, S|\theta)$ is defined, for an observation sequence y and a sequence of upper level state S, by:

$$P(y,S|\theta) = \prod_{r=1}^{R} P(S_r|S_{r-1})$$
$$\times \prod_{r=1}^{R} P(sd_i = q_r - q_{r-1}|\psi; S_{q_r})$$
$$\times \prod_{r=1}^{R} P(y_{(q_{r-1}+1, q_r]}|\phi; S_{q_r}).$$

A simple example that illustrates the modified Viterbi algorithm for HPaSMM is presented in Figure 2. For each upper level state and from beginning frame t_1 to ending frame t_N, three tables are first filled. As described in Figure 2, one is composed of the $P(y, S|\theta)$ values, the second table contains time index of the previous change of upper level state and the last table corresponds to numbering of the current upper level state. Similarly to classical Viterbi algorithm, the decoding process goes from the last frame to the first frame. The maximum $P(y, S|\theta)$ value at the last frame is selected. Then, second and third tables enable to find the current upper level state S_i and its phase beginning frame t_j. The corresponding value i is kept in $\hat{S}$ for the decoded frames (see Fig. 2). The process ends when $\hat{S}$ is completely filled, *i.e.*, when the beginning frame is reached.

2.5 A Comparison Method: Hierarchical Parallel Hidden Markov Models

The method considered for comparison purposes is based upon a hierarchical parallel hidden Markov models (HPaHMMs) build with a similar architecture than the HPaSMMs previously described. The only difference is that HPaHMMs do not model specifically upper level state durations. In a HPaHMM, the upper level state duration is supposed to follow a simple geometric law given by:

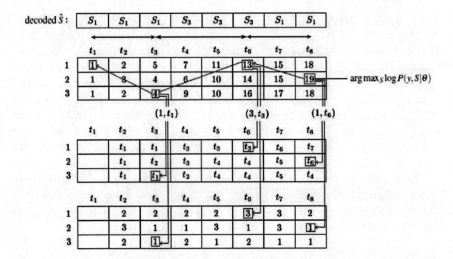

Fig. 2 Illustration of the modified Viterbi algorithm used to decode the semi-Markovian upper level sequence of activity state $\hat{S}$. This is a simple example with three upper level state S_i and eight frames T_k. Top of the image: decoded upper level state sequence $\hat{S}$ for the eight frames. For the three following tables, numbers on the vertical axis correspond to upper level states S_i numbering and the horizontal axis denotes the frames T_k. Upper table: $P(y, S|\theta)$ values. Middle table: time index of the corresponding previous change of upper level state. Down table: value of the corresponding previous upper level state. Red lines corresponds to the decoding process, from the final frame t_8 to t_1. Couples of values (S_i, t_j) between up and middle table denotes the previous upper state S_i and previous time index of upper state change t_j corresponding to current decoding frame.

$$p(d_i) = a_{ii}^{d-1}(1 - a_{ii}),$$

where a_{ii} is the probability of staying in state S_i from time t to $t+1$.

Comparison between HPaSMM and HPaHMM architectures will highlight the importance of modeling the upper level state durations provided by HPaSMMs. Fig. 3 contains the HPaHMM version of the HPaSMM presented in Fig. 1.

Hence, estimation of the set of HPaHMM parameters $\theta = \{A, \phi, \psi\}$ is similar to the HPaSMMs one except for matrix A. Indeed, since they specifically model upper level state durations, HPaSMMs training procedure computes A by looking at the transitions between activity phases only at times q_r (the time index of the end-point of the rth segment defined in the Section 2.4). For HPaHMMs training procedure, every time instant (*i.e.*, every frames) transitions of activity phases are considered to learn A. To retrieve activities using a HPaHMM, a simple Viterbi algorithm is here needed. The likelihood $P(y, S|\theta)$ to maximize is here defined, for an observation sequence y of size K and a sequence of upper level state S, by:

$$P(y,S|\theta) = \prod_{k=1}^{K} P(S_k|S_{k-1})$$
$$\times \prod_{k=1}^{K} P(y_k|\phi;S_{k-1}).$$

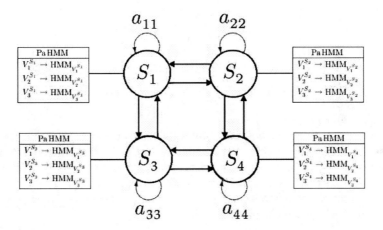

Fig. 3 Example of a HPaHMM architecture composed of $N=4$ upper level states S_i. Each state corresponds to a given phase of activity. A phase of activity is modeled by a n-layer PaHMM (one layer for each feature vector, here $n=3$).

3 HPaSMM Squash Activity Recognition

To show the efficiency of the HPaSMM framework for sports video understanding issues, an application of HPaSMM to squash video is here presented. It corresponds to a quite simple use of the HPaSMM model, with only two moving players. In Section 4, a more complex application of the novel model to handball video understanding is also proposed.

3.1 Squash Invariant Feature Representation

The proposed modeling must be able to process squash video shootings from a variety of point of view. Hence, low level activity representation should be invariant to irrelevant transformations of the trajectory data. In the video context, invariance to 2D translation, 2D rotation and scale has to be considered. In the following, we described the invariant feature values used to characterize players movements as well as their interactions.

Invariant feature values for players motion characterization

The invariant feature representation of the activity embedded in a single moving players is here presented. Following the notations introduced in Section 2.2, we consider the continuous representation of a trajectory T_k and more precisely their first and second order differential values $\dot{u}_{t,k}$, $\dot{v}_{t,k}$, $\ddot{u}_{t,k}$ and $\ddot{v}_{t,k}$. The relevant invariant representation of a single trajectory T_k is defined by the feature value $\dot{\gamma}_{t,k}$ such that:

$$\dot{\gamma}_{t,k} = \frac{\ddot{v}_{t,k}\dot{u}_{t,k} - \ddot{u}_{t,k}\dot{v}_{t,k}}{\dot{u}_{t,k}^2 + \dot{v}_{t,k}^2} = \kappa_{t,k}.w_{t,k}$$

where $\gamma_{t,k} = \arctan(\frac{\dot{v}_{t,k}}{\dot{u}_{t,k}})$ corresponds to the local orientation of the trajectory T_k, $\kappa_{t,k} = \frac{\ddot{v}_{t,k}\dot{u}_{t,k} - \ddot{u}_{t,k}\dot{v}_{t,k}}{(\dot{u}_{t,k}^2 + \dot{v}_{t,k}^2)^{\frac{3}{2}}}$ is the curvature of the trajectory T_k and $w_{t,k} = (\dot{u}_{t,k}^2 + \dot{v}_{t,k}^2)^{\frac{1}{2}}$ is the velocity of point $(u_{t,k}, v_{t,k})$.

Such a feature value characterizes both the dynamic (through velocity information) and the shape (through curvature information) of T_k. Moreover, it has been shown that the $\dot{\gamma}_{t,k}$ feature value is invariant to translation, rotation and scale in the images [9]. Hence, the feature vector used to characterize a given activity S_i of a player P_k described by the trajectory T_k is the vector containing the successive values (one for each image time index in activity S_i) of $\dot{\gamma}_{t,k}^{S_i}$:

$$\dot{\gamma}_k^{S_i} = [\dot{\gamma}_{1,k}, \dot{\gamma}_{2,k}, ..., \dot{\gamma}_{n_i-1,k}, \dot{\gamma}_{n_i,k}],$$

where n_i is the size of the processed trajectories in activity S_i.

Invariant feature values for players interaction characterization

Interaction between players is of a crucial interest in order to have an efficient representation of complex activities in sports videos. To this aim, the spatial distance (see Fig. 4) between the two players P_1 and P_2 is considered to characterize their interaction. Hence, at each successive time j, the distance between the two squash players trajectories T_1 and T_2 is defined by:

$$d_j = \sqrt{(x_{j,1} - x_{j,2})^2 + (y_{j,1} - y_{j,2})^2}.$$

More specifically, the normalized distance is computed, i.e.:

$$\tilde{d}_j = d_j / d_{norm}.$$

The feature value d_j is trivially a translation and rotation invariant feature in the 2D image plane. Moreover, to provide a scale invariant feature, a contextual normalizing factor d_{norm} has to be computed in the images. In the processed squash videos, the normalizing factor is the distance between the two sides of the court. The

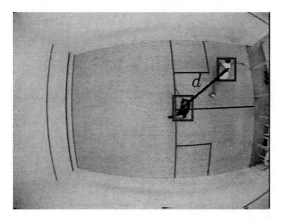

Fig. 4 A frame of a squash video. Red and blue squares correspond to players positions, d is the spatial distance between the two players.

invariant feature vector D that characterizes interaction between the two players P_1 and P_2 during activity S_i is the vector containing the successive values of $\tilde{d}_j$:

$$D^{S_i} = [\tilde{d}_1, \tilde{d}_2, ..., \tilde{d}_{n_i-1}, \tilde{d}_{n_i}],$$

where n_i is the size of the processed trajectories in activity S_i. Hence, $\dot{\gamma}_k^{S_i}$ and D^{S_i} feature vectors characterizes invariantly (to translation, rotation and scale transformations in the image plane) both the single players motions and their interaction.

3.2 Squash Activity Modeling Using HPaSMM

This section describes the HPaSMM used to process squash videos. As illustrated by Figure 5, it relies on two upper level states S_1 and S_2, and on the three feature values defined in the Section 3.1.

Upper layer: Activity modeling by semi-Markov chains

Upper level states S_1 and S_2 of the squash HPaSMM define "rally" and "passive" activity phases. In the illustration in Figure 5, the upper level layer is surrounded in red and is composed of two upper level states S_1 and S_2.

A is the HPaSMM state transition probability matrix at $\{q_i\}$ the time index of the end-point of the ith segment (see Section 2.4). In the proposed modeling with only two upper level states, $a_{21} = a_{12} = 1$. Indeed, when a segment of activity ends, the system necessarily passes from one upper level state to the other one (i.e., from S_1 to S_2 or from S_2 to S_1).

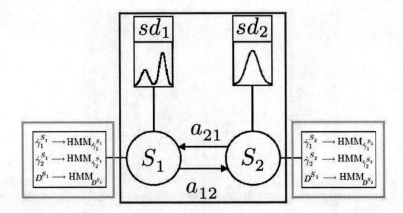

Fig. 5 Squash HPaSMM with two upper level states corresponding to activity phases S_1 and S_2. Each of these upper level states is characterized by three HMMs (modeling respectively $\dot{\gamma}_1^{S_i}$, $\dot{\gamma}_2^{S_i}$ and $\tilde{d}^{S_i}$ for the considered state S_i) and by a GMM modeling sd_i (i.e., the state duration density in the upper level state S_i). Upper level layer is surrounded by red square and lower level layer is surrounded by green squares.

Lower layer: feature modeling using parallel hidden Markov models

Tackling with the lower layer modeling, the PaHMM modeling proposed in [25] has been used to build a probabilistic modeling of the spatio-temporal behavior of the feature vectors $\dot{\gamma}_1^{S_i}$, $\dot{\gamma}_2^{S_i}$ and D^{S_i}. Figure 5 presents the lower level layer surrounded in green and composed of 3-layer PaHMMs.

3.3 Considered Data Sets and Experiments

The proposed temporal segmentation framework has been tested on trajectories from squash videos taken from the "CVBASE'06" sports video database [3]. It provides squash videos as well as trajectories of the squash players in the images. Game activity phases ("rally" and "passive" phases) that we use as ground truth for experimental evaluations are also available. An illustration is given in Figure 6 which shows two frames from the processed squash video. According to ground truth, the left one corresponds to "rally" activity and the right one to "passive" activity.

Trajectories of the first half of the squash video (corresponding to 7422 frames) were used for training a HPaSMM with two states S_1 and S_2 corresponding to activity phases "rally" and "passive". Trajectories of the second half (corresponding to 8086 frames) of the video were used to test the proposed modeling. Fig. 7 presents the squash players trajectories of the squash videos respectively used for training and testing.

The activity phase denoted by "rally" is defined by the period between the beginning of a point (a player serves) and its end (when ball bounces twice or when a

Understanding Sports Video Using Players Trajectories 137

Fig. 6 Two frames taken from the squash video used for experiments (the entire video is composed of 15508 frames). Ground truth activity for left and right images are respectively "rally" and "passive".

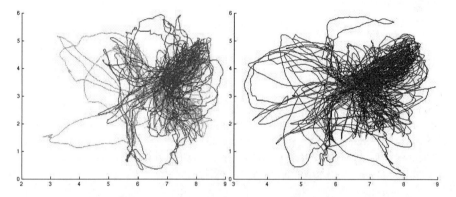

Fig. 7 Left: training trajectories of 2 squash players in the video plane (in green and magenta) during the first half (corresponding to 7422 images) of a squash video. Right: test trajectories of 2 squash players (in blue and red) during the second half of the squash video (corresponding to 8086 images).

player hits the ball out). The "passive" phase corresponds to periods between the end of a point and the beginning of the following point, and contains (despite its name) an important activity. Indeed between two points, one of the player has to reach and take the ball and then players may change positions whether the server won the point or not. Furthermore, on the contrary to beginners that do not have efficient displacements, professional players (such that those in the processed squash video) usually only have low motion activity during the "rally" phase. Position of the players are more important than their mobility, *i.e.*, they place themselves to optimize their displacements in order to easily and quickly reach the ball while keeping from tiredness.

Hence, it is difficult to visually determine if the two squash players are in a "rally" phase or in a "passive" phase in the video. Players movements may be very reduced both in the "rally" phase and in the "passive" phase. Indeed, even during the video segments of "rally" activity, there are periods during which players are almost statics, so that it looks more like a "passive" phase. To visually proceed a temporal segmentation, a simpler way seems to focus on the relative distance evolution between players (*i.e.*, the evolution of the feature $\tilde{d}$). Hence, such experiments are of interest and may justify the chosen activity characterization that relies on players motions and on their interaction.

In the sequel, we will evaluate the performance of the method as the ratio of correctly classified images (with respect to the "rally" and "passive" activity phases) and the total number of processed images. To this end, ground truth on the entire set of trajectories is exploited. All the reported results were obtained using k_{group} and h parameter values respectively equals to 8 and 1 (as defined in Section 2.2).

3.4 Results

Hence, the first 7422 frames of the video (and more precisely the corresponding trajectories) where used to learn squash HPaSMM showed in Fig. 5. This step results in the computation of the parameter set $\theta = \{A', \phi, \psi\}$ defined in Section 2.3. Fig. 8 shows the training result when fitting a GMM on the duration state distribution of respectively the "rally" and "passive" upper level state.

A 89.2% of correct classified images has been reached on the 8086 testing images with the squash HPaSMM. Corresponding results are presented in the upper part of Figure 9. Using the comparison Markovian method (*i.e.*, the squash HPaHMM model developed in Section 2.5), a 88% correct recognition rate was obtained. Corresponding results are shown in lower part of figure 9.

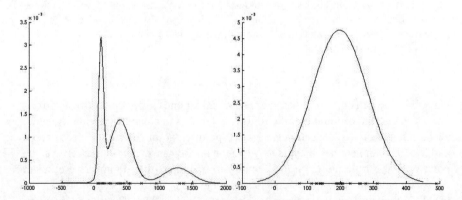

Fig. 8 Left: duration state density modeling using GMM for "rally" upper level state. The x-axis corresponds to the observed state durations (crosses indicate observed state durations). Left : duration state density modeling using GMM for "passive" upper level state.

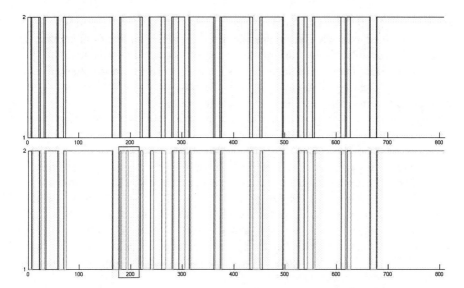

Fig. 9 Up: Ground truth is plotted in red. Result of the HPaSMM processed temporal segmentation is superimposed in blue. The x-axis corresponds to the frame index. The "1" and "2" values respectively correspond to the "passive" and "rally" phases. Down: Ground truth is plotted in red. Result of the HPaHMM processed temporal segmentation is superimposed in green. The x-axis corresponds to the frame index. The "1" and "2" values respectively correspond to the "passive" and "rally" phases. The purple surrounded region contains the main difference between results using HPaSMM and HPaHMM.

Experimentations have also been performed without taking into account the feature vectors D^{S_i} describing the evolution of the distance between the two players. Hence, only $V_1^{S_i}$ and $V_2^{S_i}$ feature vectors were here considered. Recognition results, for the same experimental training and testing sets, fall from 89.2% to less than 70%.

Finally, the HPaSMM method has been tested based only on interactions. This means that activity phases characterization only relied on the distance feature vectors D^{S_i}, and not on feature vectors $V_1^{S_i}$ et $V_2^{S_i}$. Recognition results using only interactions between players reaches a 86.8% positive rate. This is a little fewer than the 89.2% rate obtained when considering interactions as well as players movements.

A recognition rate of 89.2% seems a satisfying results. Moreover, the HPaSMM algorithm retrieved the exact number of activity phases. The rate of 10.8% recognition errors only corresponds to lags between the detected phase changes. Hence, every transition between "rally" and "passive" activity phases have been detected, retrieving the exact number of played points (*i.e.*, 13 points).

Comparison between HPaSMM and HPaHMM (with state durations not modeled by GMMs, *i.e.* state duration in state i following a geometric law in regards of a'_{ii}) is also of interest. Most of the time, HPaHMM gave a quite similar segmentation.

However, HPaSMM provides an imperfect segmentation. Indeed, 14 points instead 13 were detected, hence retrieving one additional (non-existing) point. It highlights the contribution given by the semi-Markov upper level modeling, validating the hypothesis that state duration in upper level state does not follow a geometrical law and needs a more adapted modeling. This is illustrates in figure 9, were the purple surrounded region contains the false "rally" detection.

Results below 70% of good phase segmentation obtained when not considering the $\tilde{d}$ feature value show the decisive inherent information of the temporal trends of the distance (*i.e.*, of the interactions). So, single players motions are not sufficient to understand the observed activities. Results obtained using only the interaction feature vectors D^{S_i} shows that the interactions between players is a crucial information for activity phase retrieving. Moreover, comparison with HPaSMM method when using D^{S_i} as well as $V_1^{S_i}$ and $V_2^{S_i}$ feature vectors shows that exploiting dynamics and shapes of the single trajectories in addition to interactions helps getting a better recognition. These results validates the hypothesis made by our modeling that both single players dynamics and their interactions are to be considered to understand activities in sports such that squash.

Computation time

We give here computation time results using an Intel Pentium Centrino 1.86 GHz processor. Learning the set of HPaSMM parameter $\theta = \{A', \phi, \psi\}$ with the 7422 frames training set took around 1 minute long. Computation time for the recognition stage on the 8086 frames was around three minutes. The HPaHMM computation time is slightly lower than the HPaSMM one.

Nevertheless, when considering HPaSMM only with the feature vectors D^{S_i} used for activity representation (results skipping from 89.2% to 86.8%), computation times become much lower. Indeed, most of the squash HPaSMM computation time is spent computing the continuous representation necessary to get the $\dot{\gamma}$ values. Hence, when considering only the D^{S_i} feature vectors, no kernel approximation is further needed and the computation times are much smaller (divided by around 10 times).

4 HPaSMM Handball Activity Recognition

In order to show the efficiency of the HPaSMM framework for sports video understanding issues, an application of HPaSMM to handball video is here presented. It is a more complex use of the HPaSMM model than with squash videos, since handball teams are formed with seven players and the diversity of activities is higher.

4.1 Handball Invariant Feature Representation

To process semantic segmentation of handball videos, motions in the court plane (*i.e.*, the trajectories) of the players of a single team are considered. Similarly to

Understanding Sports Video Using Players Trajectories

squash feature values (see Section 3), the choice of the feature representation of handball team activities is based upon the single dynamics of players as well as their interactions. Three main ideas led the choice of the feature values:

- to consider the global dynamics of fielders (*i.e.*, any handball player except the goalkeeper),
- to exploit the specific role and location of the goalkeeper,
- to keep a limited number of feature values.

Invariant feature values for players motion characterization

To characterize players movements and to handle the first and the third ideas defined above, *i.e.* to take into account the global dynamics of fielders within a limited number of feature values, three feature values are computed at each time instant t. These feature values are the min, mean and max values of distance between successive positions (*i.e.*, the distance covered between $t-1$ et t) of the six fielders, respectively denoted by $d_{intramin,t}$, $d_{intramean,t}$ and $d_{intramax,t}$. These three feature values gives a reduce yet important and global information on the dynamics of fielders. Computed distances correspond to Euclidean distances between trajectories in the court plane. The values $d_{intramin,t}$ for an activity S_i are gathered in the vector $D^{S_i}_{intramin}$:

$$D^{S_i}_{intramin} = [d_{intramin,1}, ..., d_{intramin,n_i-1}, d_{intramin,n_i}],$$

where n_i is the size of the processed trajectories in activity S_i. The same holds for the two other feature values $d_{intramean,t}$ and $d_{intramax,t}$ to produce vectors $D_{intramean}$ and $D_{intramax}$.

Invariant feature values for players interaction characterization

To characterize players interaction while handling the second and the third ideas outlined in the introduction of this section, *i.e.* to exploit the specific role of the goalkeeper within a limited number of feature values, two feature values are computed at each time instant t:

- the mean distance between the goalkeeper and the six fielders $d_{GC,t}$,
- the mean distance between the six fielders $d_{C,t}$.

The values $d_{GC,t}$ for an activity S_i are gathered in the vector $D^{S_i}_{GC}$:

$$D^{S_i}_{GC} = [d_{GC,1}, ..., d_{GC,n-1}, d_{GC,n}],$$

where n_i is the size of the processed trajectories in activity S_i. The same holds for the feature values $d_{C,t}$ to produce feature vectors $D^{S_i}_C$.

It is important to precise that, on the contrary to squash trajectories feature representation, no kernel approximation is here needed. Indeed, the five considered feature values do not require first or second differential values and are directly computed on the row trajectory data, *i.e.*, on $T_k = \{(x_{1,k}, y_{1,k}), ..., (x_{l_k,k}, y_{l_k,k})\}$.

4.2 Handball Activity Modeling Using HPaSMM

This section describes the HPaSMM used to process squash videos. As illustrated by Figure 10, it relies on eight upper level states and on the five feature values defined in the Section 4.1.

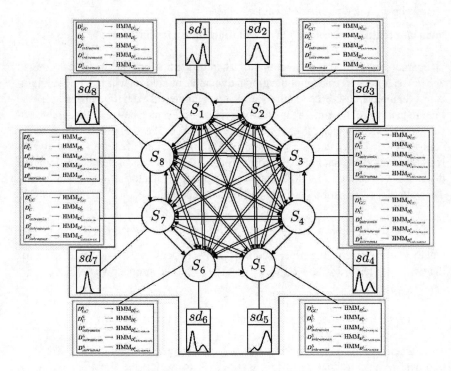

Fig. 10 HPaSMM modeling of handball composed of eight upper level states S_i. Each state corresponds to a given phase of activity. A phase of activity is modeled by a five-layer PaHMM (one layer for each feature vector) and by a GMM modeling the state durations (sd_i is the state duration associated to state S_i). Upper level layer is surrounded by a red square and lower level layer is surrounded by green squares. For display reasons, feature vectors $D_{GC}^{S_i}$ are denoted by D_{GC}^i, and the same holds for feature vectors $D_C^{S_i}$, $D_{intramin}^{S_i}$, $D_{intramean}^{S_i}$ and $D_{intramax}^{S_i}$.

Upper layer: Activity modeling by semi-Markov chains

Handball activity phases are defined by HPaSMM upper level states S_i. We consider here eight phases of activity, defined along with their numbering as follows:

- "slowly going into attack": activity phase 1,
- "attack against set-up defense": activity phase 2,

- "offense free-throw or timeout": activity phase 3,
- "counterattack, fast break": activity phase 4,
- "returning, preventing from fast break": activity phase 5,
- "slowly returning in defense": activity phase 6,
- "defense": activity phase 7,
- "defense free-throw or timeout": activity phase 8.

Such a variety of activity provides a quite complete picture of the possible handball activities, providing an exhaustive representation for efficient understanding of handball videos. In the illustration in Figure 10, the upper level layer is surrounded in red and is composed of eight upper level states.

Lower layer: feature modelings using parallel hidden Markov models

The five feature values that describes both dynamics of the trajectories and interactions between trajectories, *i.e.*, $d_{GF,t}$, $d_{F,t}$, $d_{intramin,t}$, $d_{intramean,t}$, and $d_{intramax,t}$ are used to characterize the eight phases of activity S_i. For each upper level state, the five corresponding feature vectors are modeled using a 5-layer PaHMMs. Figure 10 presents the lower level layer surrounded in green.

4.3 Integrating Audio Information: Recognition of Referees Whistles

To facilitate recognition of activity phases and more precisely their transitions, audio information is taken into account. Indeed, audio data contains an important information to specify transitions between activities: referees whistles.

To retrieve referees whistles instances, the audio stream is processed using two free access softwares: Spro and Audioseg, available online [26, 2]. Spro provides a description of whistles request and of the processed audio stream contained in the handball video. This characterization is based upon cepstral coefficients of mel frequency [15] computed in successive time intervals defined by a sliding window. Audioseg then performs recognition of the request within the audio stream by comparing the coefficients of each audio intervals using a Dynamic Time Warping procedure [16].

Detection of referees whistles is integrated in the HPaSMM method in a simple way: each whistle corresponds to an activity phase change in the model. Hence, in the Viterbi decoding algorithm, the hypothesis is made that the current activity phase is stopped each times a whistle is detected and then another phase begins.

Hence, a partition of the observed actions into successive segments Seg_k is given by referees whistles, where a segment is included between two referees whistles. Each segment can then be decoded separately using the Viterbi algorithm. Since each whistle corresponds to a change of activity, decoding of successive video segments is simply made by specifying that first activity phase of a segment Seg_{l+1} has to be different from the last activity phase found for the previous segment Seg_l.

Setting-up of a leave-one-out cross validation method

The first advantage in taking into account referees whistles is to help recognition of some transitions between activity phases. Moreover, it facilitates establishment of a leave-one-out cross validation (LOOCV) test method. Indeed, the set of trajectory is "cut" into $N_w + 1$ segments (defined by N_w detected whistles). Viterbi decoding of one segment is computed independently from the other segments, the only information to integrate being that the first activity phase of a segment Seg_{k+1} has to be different of the last activity phase of the segment Seg_k. Hence, a LOOCV method is exploited. Each segment is decoded independently while the N_w other segments are used for training HPaSMM as well as HPaHMM models. In the following, this procedure will allow the use of larger training sets.

4.4 Considered Data Sets and Experiments

The proposed method has been tested on a set of trajectories of handball players belonging to a same team. The trajectory database is available online [3]. To extract these trajectories, two bird-eye view cameras were used, one above each half of the court plane. To this aim, a modified color-histogram-based condensation tracking method was computed on the video sequences [14, 21]. Database providers supervised the tracking procedure by correcting errors that appeared during the process. An appropriate calibration step was also introduced to map the image coordinates in the court plane and to compensate for the observed radial distortion. An estimation of the error on the players coordinates in the court between 0.3 and 0.5 m has been obtained [20]. For each player, 25 coordinate data (one per frame) per second are available. To reduce the tracking jitter while preserving the measurement accuracy, a Gaussian smoothing was finally applied.

Three images corresponding to a same given time instant of a handball match are provided in Figure 11. The two images on the left were extracted from videos acquired by bird-eye view cameras. The right one is a frame of a video shot by a hand-held camera. Figure 12 contains the entire set of trajectories of the handball

Fig. 11 Three images of a same instant in three videos of a given handball action. The two left images have been extracted from two bird-eye view videos, the third one from a hand-held camera.

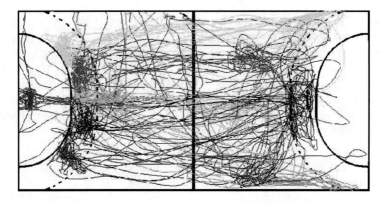

Fig. 12 Complete set of available trajectories of the seven players belonging to a same handball team (each trajectory is plotted in a given color and corresponds to a different player). It corresponds to about ten-minutes of video (more precisely, 14664 frames).

players belonging to a same handball team. They correspond to approximately ten minutes of video (14664 frames).

The database also provides a ground truth annotation of the video within several activities. The set of activities considered here is slightly different from the eight activities we have chosen. Indeed, thanks to ground truth, activities denoted by "7-meter throw" and "Jump ball at the court center" are observed only once in the processed ten minutes video. Hence, in order to generate ground truth, we were forced to segment these events using the eight defined activities. With a larger amount of trajectories, such events could have been properly modeled. This stresses the importance of having a sufficient amount of training data adapted with the considered modeling.

Likewise, the event "Back in the court center" (after conceding a goal) is observed only twice in the whole video. This activity would also have required the setting-up of an additional upper state. Again, it was not possible to efficiently model it due to the available trajectory data sets. In order to produce corresponding ground truth, we had to resort to the activities "defense free-throw or timeout" and "slowly going into attack" which depict the best a similar motion content.

Hence, we used the available annotation to construct our activity ground-truth with the eight activity phases defined in Section 4.2. Provided ground truth also allowed us to validate the hypothesis that every whistle induces a change in the activity phases.

The experiments carried out to test the proposed HPaSMM and HPaHMM methods for trajectory-based handball activity recognition are now described.

In our experiments, data sets respectively used for training and testing the proposed algorithm correspond to distinct periods of the ten minutes handball video. Hence, no overlapping data were used for training and testing. However, due to the available trajectory data sets, training and testing sets of trajectories are video sequences of the same team in a same game. It would be an add to test the proposed method on trajectory data from other game and other teams.

The first experiments processed use of the HPaSMM and HPaHMM methods without considering audio information. A part of the handball players trajectories is used for training the HPaSMM and HPaHMM models. It includes 6370 images, *i.e.*, more than 4 minutes of video. A second part of the trajectory set is then used for testing and comprises 8294 images, *i.e.*, little less than 6 minutes of video. Figure 13 presents test and training data sets.

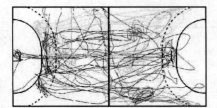 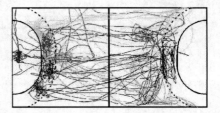

Fig. 13 Left: training set of handball players trajectory (6370 frames). Right: testing set of handball players trajectory (8294 frames).

We have also tested the proposed HPaSMM method while taking into account audio detection of the referees whistles. The LOOCV validation method was then considered, allowing larger training data sets while keeping non-overlapping test and training data sets (see Section 4.3).

Observations from other video streams have also been included to train the upper level state transition matrix A and the GMMs modeling the state duration sd_i. Indeed, trajectories are not required to train this subset of parameters. Hence, other handball videos were used and manually segmented into the eight considered activity, hence providing additional information in term of transitions between activity phases and of durations of activity phases. These handball activity data have been extracted from 2008 Beijing Olympic Games handball final. The videos are available online [5].

In the sequel, we will evaluate the performance of the method as the ratio of correctly classified images (with respect to the eight defined activity phases) and the total number of processed images. To this end, ground truth on the entire set of trajectories is exploited. All the reported results were obtained using k_{group} parameter value equal to 8 (as defined in Section 2.2).

4.5 Results

We report here results corresponding to the experiments described in the previous section. First, when using training and testing data sets of respectively 6370 and 8294 frames with the HPaSMM method, a rate of 76.1% correct recognition was obtained. Results are plotted in Figure 14 which contains the ground truth and the

Understanding Sports Video Using Players Trajectories

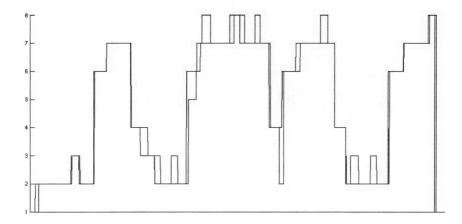

Fig. 14 Plot of the recognition results obtained by the HPaSMM method. First part of the trajectory set (6370 images, see Fig. 13) was used for training while the second part (8294 images, see Fig. 13) was kept for recognition test. Ground truth is plotted in red and recognition results are plotted in blue. Numbers on the vertical axis correspond to the numbering of activity phases described in Section 4.2. The horizontal axis denotes the numbering of the successive groups of images (with parameter k_{group} value set to 8, see Section 2.2).

segmentation results. With the HPaHMM method and the same training and testing data sets, a rate of 72.7% has been reached.

Such results can be explained by the lack of training trajectories. For sure, 4 minutes for the training set does not seem sufficient to efficiently train such "complex" models. Moreover, most of the errors occurs for the offense and defense free-throw or timeout activity phases. To alleviate these shortcomings, we will now exploit audio information contained in the audio stream. Indeed, using the LOOCV available when processing audio information will provide us with larger training sets while helping detection of transitions of free-throw or timeout activity phases.

Hence, in the following, we first present results of referees whistles extraction. Then, results of activity understanding conducted with the leave-one-out cross validation method, which enables to consider more training data, are reported.

Results of whistle audio recognition

The method presented in Section 4.3 for recognition of referees whistles gave very satisfying results. Indeed, 29 referees whistles out of 31 where recognized while no false detection was detected. Figure 15 illustrates these results, where green lines correspond to detected whistles and brown ones are undetected whistles.

We now use the referees whistles (29 referees whistles were correctly detected in the ten-minutes audio stream) in the HPaSMM and HPaHMM methods.

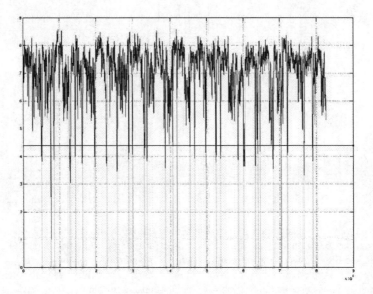

Fig. 15 Results of referees whistles recognition obtained on the ten minutes of audio stream. Audio stream is plotted in blue, detection threshold is in red. The 29 detected whistles are represented by green lines while undetected ones are plotted with brown lines.

Results using audio information and LOOCV method for HPaSMM

When the LOOCV method is applied to the 30 segments of handball video trajectories, a performance of 89.8% correct activity recognition was reached. However, recognition errors mainly occur when processing events "7-meter throw" and "Jump ball at the court center" (see Section 4.4). Then, we omitted these two activities (*i.e.*, 5 of the 30 video segments). 87% of the whole ten-minutes video was then concerned, that is more than 8 minutes 30 seconds, and a correct recognition rate of 92.2% was obtained.

The complete set of activity phase has been correctly recovered, while the 7.8% of errors corresponds to time-lags at the transitions between activities. Most of these time-lags are around one or two seconds long. Hence, as soon as the training data set is adapted to the activity architecture modeled in the upper layer of the model, the method supplies a quite satisfying understanding of the observed video.

Figure 16 and Table 1 illustrate the recognition results obtained with the HPaSMM after discarding the five segments corresponding to events "7-meter throw" and "Jump ball at the court center". Table 1 contains, for each activity phase, the correct recognition rates.

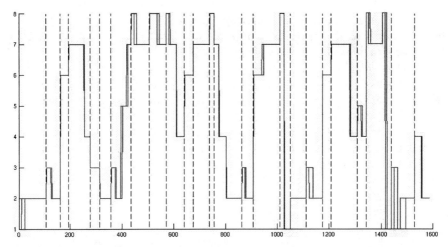

Fig. 16 Plot of the recognition results obtained by the HPaSMM method with the LOOCV procedure. Ground truth is drawn in red while the recognition results are plotted in blue. Detected referees whistles are indicated by dotted lines. Numbers on the vertical axis correspond to activity phases numbering as described in Section 4.2. The horizontal axis denotes the numbering of the successive group of images (with parameter k_{group} value set to 8).

Table 1 Correct recognition rates obtained by the HPaSMM method for the eight activity phases. Experiments were conducted using the LOOCV test method, without considering the events "7-meter throw" and "Jump ball at the court center". The number of images corresponding to each activity phase is also indicated.

	Recognition rate	# images
Slowly going into attack	68.1	728
Attack against set-up defense	92.4	3776
Offense free-throw or timeout	87.1	1056
counterattack, fast break	95.6	1280
Returning, preventing from fast break	88.6	280
Slowly returning in defense	93.6	1112
Defense	100	3592
Defense free-throw or timeout	81	928

Comparison with the HPaHMM method using audio information and LOOCV

Referees whistles were integrated in the HPaHMM method in a similar way than with the HPaSMM one. Hence, each segment is decoded separately and the LOOCV test method defined above is also considered with HPaHMM method.

We report here the result of the HPaHMM method when processing the set of trajectories of about 8 minutes and 30 seconds (still omitting the "7-meter throw" and "Jump ball at the court center" events). A correct recognition rate of 89.8% was here obtained with the HPaHMM. This result has to be compared with the 92.2% correct recognition rate obtained with the HPaSMM method.

For a large majority of the 25 processed video segments, HPaHMM gave similar segmentations than those obtained with the HPaSMM methods. However, for some segments, the HPaHMM method provided a less accurate segmentation than the HPaSMM method. Figure 17 illustrates these results. It presents results obtained by both methods on the first segment of the processed handball video. For that segment, correctness rate is 60% with HPaHMM method whereas the HPaSMM one reaches 89% correct recognition rate.

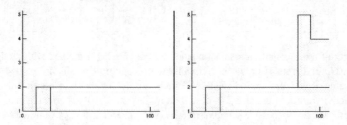

Fig. 17 Plot of the recognition results for the first segment of the video. Left: results obtained with the HPaSMM method. Right: results obtained with the HPaHMM method. For both methods, ground truth is plotted in red and obtained results are plotted in blue. Numbers on the vertical axis correspond to activity phases numbering described in Section 4.2. The horizontal axis denotes the numbering of the successive group of images (with parameter k_{group} value set to 8).

Such results highlight the advantage of modeling upper state durations since the HPaSMM method obtains more efficient recognition results than the HPaHMM one. First, this information enables to reduce time shifts that may appear at the transitions between activity phases. Furthermore, as shown on Figure 17, this information also helps HPaSMM method to avoid errors of activity recognition that may occur with the HPaHMM method.

Computation time

For the conducted experiments, we used a Intel Pentium Centrino 1.86 GHz processor. k_{group} was set to 8 (see Section 4.1). Computation times reported here correspond to tests with the LOOCV procedure when omitting events "7-meter throw" and "Jump ball at the court center". Hence, training set of trajectories was composed of 24 segments corresponding to approximately 8 minutes and 30 seconds depending on the size of the segment left out and used for testing.

The computation time required to train the HPaSMM model was around 3 seconds. Computation time for the recognition stage applied to one segment (the mean size of a segment is 20 seconds) is around 6 seconds. Hence, the overall activity retrieval process took 2 minutes and 20 seconds. The HPaHMM computation time is a little lower than the HPaSMM one. Computation times are much smaller than those reported for squash video processing (see Section 3.4). Indeed, kernel approximation of the trajectories is the procedure that took most of the computation time for squash video processing and it is not needed here.

Moreover, since referees whistles detection allows to decode each segment independently, it is possible to decode a segment as soon as it ends. When a whistle is blown and detected, around six seconds (depending on the size of the processed segment) are necessary to retrieve activity phases in that last segment. Using audio information hence allows an almost real-time processing of handball trajectories.

5 Conclusions, Extensions and Perspectives

We described here an original hierarchical semi-Markovian framework for sports understanding purposes that uses trajectories extracted from videos. An upper level layer is dedicated to modeling of high semantic activity while a lower level layer is used to process low level feature values. This is a general method that may be used for many kind of sports, and more specifically for team and racquet sports. Trajectories of players reconstructed in the court plane are used for activity phases recognition.

The proposed method has been adapted to two very different sports: squash and handball. For each of these two sports, a specific representation taking into account both dynamics and interaction information is proposed and a set of relevant activity has been defined. The developed HPaSMM method has been tested on two sets of trajectories corresponding respectively to approximately 10 minutes of a squash game and of a handball game. Experiments gave very satisfying results, showing efficiency and easiness of the adaptation of the proposed modeling to very different sports. Comparison with Markovian modeling highlighted the importance of modeling activity state durations provided by semi-Markovian modeling. Extensions to more complex model architectures could be handled with larger sets of trajectories.

Thanks to the explicit activity phase modeling, extensions of the proposed method to other sports (such that soccer, basket-ball, tennis, volley-ball) looks straightforward. To this aim, at a lower level and depending on the processed sport, a specific feature value representation would have to be defined. However, it would still be important to adapt a reduce feature value representation that keeps taking into account both for individual players movements and for interaction. Moreover, activity phases definition is depending on the considered sport, and also on the desired degree of automatic understanding.

Based on these modelings, interesting applications would be considered in several domains. Firstly, relevant use for the mass media are reachable, including for example automatic sports video summarization and video on demand requests. Besides, such semantic analysis are of great interest in the sports professional field, allowing automatic players statistics generation, tactical phases understanding...

Finally, perspectives of extension to "general" high level analysis of videos may also be reachable, especially by extending the proposed scheme for activity-based processing of sports videos to video surveillance (for example, crowd event analysis, steal detection, dropping off suspect object recognition...). The use of scenario definition tools (for example, the logical grammars recently investigated in [24]) would be beneficial, providing ways to create complex and extended scenario into a statistical framework such that the one presented in this work.

References

1. Anjum, N., Cavallaro, A.: Multifeature object trajectory clustering for video analysis. IEEE Trans. on Circuit and Systems for Video Technology, Special issue on event analysis in videos 18(11), 1555–1564 (2008)
2. http://gforge.inria.fr/projects/audioseg
3. http://vision.fe.uni-lj.si/cvbase06/download/dataset.html
4. http://vision.fe.uni-lj.si/cvbase06/download/CVBASE-06manual.pdf
5. http://www.dailymotion.com/
6. Ge, X.: Segmental semi-Markov models and applications to sequence analysis. Ph. D. thesis. University of California (2002)
7. Günsel, B., Tekalp, A.M., van Beek, P.J.L.: Content-based access to video objects: temporal segmentation, visual summarization, and feature extraction. Signal Processing 66(2), 261–280 (1998)
8. Hervieu, A., Bouthemy, P., Le Cadre, J.-P.: A HMM-based method for recognizing dynamic video contents from trajectories. In: Proc. of the IEEE International Conference on Image Processing (2007)
9. Hervieu, A., Bouthemy, P., Le Cadre, J.-P.: A statistical video content recognition method using invariant features on object trajectories. IEEE Trans. on Circuit and Systems for Video Technology, Special issue on event analysis in videos 18(11), 1533–1543 (2008)
10. Hervieu, A., Bouthemy, P., Le Cadre, J.-P.: Trajectory-based handball video understanding. In: Proc. of the ACM Conference on Image and Video Retrieval (2009)
11. Hongeng, S., Nevatia, R., Bremond, F.: Large-scale event detection using semi-hidden Markov models. In: Proc. of the IEEE International Conference on Computer Vision (2003)
12. Jung, C.R., Hennemann, L., Musse, S.R.: Event detection using trajectory clustering and 4-D histograms. IEEE Trans. on Circuit and Systems for Video Technology, Special issue on event analysis in videos 18(11), 1565–1574 (2008)
13. Kokaram, A., Rea, N., Dahyot, R., Tekalp, M., Bouthemy, P., Gros, P., Sezan, I.: Browsing sports video (Trends in sports-related indexing and retrieval work). IEEE Signal Processing Magazine 23(2), 47–58 (2006)
14. Kristan, M., Perš, J., Perše, M., Bon, M., Kovačič, S.: Multiple interacting targets tracking with application to team sports. In: Proc. off the International Symposium on Image and Signal Processing and Analysis (2005)
15. Mermelstein, P.: Distance measures for speech recognition, psychological and instrumental. In: Proc. of the Workshop on Pattern Recognition and Artificial Intelligence (1976)
16. Muscariello, A., Gravier, G., Bimbot, F.: Variability tolerant audio motif discovery. In: Proc. of the International Conference on Multimedia Modeling (2009)

17. Natarajan, P., Nevatia, R.: Coupled hidden semi-Markov models for activity recognition. In: Proc. of the IEEE International Joint Conference on Artificial Intelligence (2007)
18. Natarajan, P., Nevatia, R.: Hierarchical multi-channel hidden semi-Markov models. In: Proc. of the IEEE Workshop on Motion and Video Computing (2007)
19. Oliver, N.M., Rosario, B., Pentland, A.P.: A Bayesian computer vision system for modeling human interactions. IEEE Transactions Pattern Analysis and Machine Intelligence 22(8), 831–843 (2000)
20. Perš, J., Kovačič, S.: Tracking people in sport: making use of partially controled environment. In: Proc. of the International Conference on Computer Analysis of Images and Patterns (2001)
21. Perše, M., Perš, J., Kristan, M., Vučkovič, G., Kovačič, S.: Physics-based modeling of human motion using Kalman filter and collision avoidance algorithm. In: Proc. of the International Symposium on Image and Signal Processing and Analysis (2005)
22. Perše, M., Kristan, M., Kovačič, S., Vučkovič, G., Perš, J.: A trajectory-based analysis of coordinated team activity in a basketball game. Computer Vision and Image Understanding 113(5), 612–621 (2009)
23. Piciarelli, C., Micheloni, C., Foresti, G.L.: Trajectory-based anomalous event detection. IEEE Trans. on Circuit and Systems for Video Technology, Special issue on event analysis in videos 18(11), 1544–1554 (2008)
24. Richardson, M., Domingos, P.: Markov logic networks. Machine Learning 62, 107–136 (2006)
25. Vogler, C., Metaxas, D.: Parallel hidden Markov models for American sign language recognition. In: Proc. of the IEEE International Conference on Computer Vision (1999)
26. http://gforge.inria.fr/projects/spro

Real-Time Face Recognition from Surveillance Video

Michael Davis, Stefan Popov, and Cristina Surlea

Abstract. This chapter describes an experimental system for the recognition of human faces from surveillance video. In surveillance applications, the system must be robust to changes in illumination, scale, pose and expression. The system must also be able to perform detection and recognition rapidly in real time.

Our system detects faces using the Viola-Jones face detector, then extracts local features to build a shape-based feature vector. The feature vector is constructed from ratios of lengths and differences in tangents of angles, so as to be robust to changes in scale and rotations in-plane and out-of-plane. Consideration was given to improving the performance and accuracy of both the detection and recognition steps.

1 Introduction

CCTV cameras have become ubiquitous, nowhere more so than in the United Kingdom. So far, the value of CCTV surveillance has failed to live up to the hype. A recent report by the Metropolitan Police in London revealed that the city's one million-plus cameras have helped to solve only a handful of crimes [18]. The criminologist Clive Norris points out [21] that CCTV operators tend to act on their prejudices (for example, focussing cameras on people because of their skin colour) or merely on scenes which they find entertaining, to relieve the boredom of staring at mundane street scenes all day. So far, the main CCTV success stories are for forensic use[1] and as a deterrent against some forms of petty or opportunistic crime[2].

Michael Davis · Stefan Popov · Cristina Surlea
School of Electronics, Electrical Engineering and Computer Science
Queens University Belfast, Belfast, BT7 1NN, UK
e-mail: `mdavis05@qub.ac.uk`

[1] See [25] for a recent example.
[2] As predicted by Jeremy Bentham's *Panopticon* [4], people change their behaviour when they know they are under surveillance. This is not only true of criminals. For example, ordinary people change their behaviour within airports to flow through the system of control [22].

CCTV surveillance is often justified on the basis that it can be used to prevent terrorist attacks, but the reality is somewhat more banal. As [26] points out, those committing acts of terrorism are usually unknown before the act takes place, and it is very difficult to conduct surveillance of someone whose identity is unknown. There are few examples of CCTV surveillance being used effectively for real-time detection or prevention of crime.

It has been suggested that the effectiveness of CCTV surveillance can be improved by tracking known individuals as they move through a surveillance network. One possible application could be to follow known hooligans as they enter and move about a football stadium. This paper will consider the design of a video surveillance network which could track the movement of known individuals in real time.

1.1 Real-Time Video Surveillance

Let us consider a specific application area — video surveillance within a Secure Corridor; for example, within an airport or other secure building. Airports can be considered as a self-contained microcosm, which can be viewed in terms of the flows through them, and the movements and behaviours that take place in their arrival, shopping and transit areas [15].

The airport can also be viewed as a *Surveillant Assemblage*. The relationship between surveillance and mobility within an airport is considered in [22]. Until recently, airport surveillance did not focus on individuals – people were disassembled into data flows to monitor movement through the airport. Individual surveillance has come to the fore within the last ten years. Passenger profiling is conducted ostensibly for security, but profiling information is also of commercial use to the airport authorities. The combination of profiling with the ability to track the movement of individuals is potent.

A number of airports have already installed systems to register passengers as they enter the Secure Corridor, and track when they leave. Passengers are enrolled using their boarding card and a digital photograph is taken. When the passenger leaves the Secure Corridor, the photograph is compared with the passenger's face by a security officer. This generates a transaction when the passenger enters and leaves the secure area, but cannot track their movements within it. Airports may wish to track some individual passengers as they move through passport control, the retail shopping area, departure lounges and exit to their gate. Our hypothetical system could achieve this by tracking the passenger's face as they move around the secure area.

The proposed system will need to work in real time: that is, the extraction of facial features and comparison to the stored model for each passenger must be extremely fast. The metric used must be reasonably robust to changes in scale, illumination, pose and occlusion. The system will need to track individuals as they move from the field of view of one camera to another. A distributed network of cameras will need to communicate with a back-end database and pattern-matching system. These requirements are discussed in more detail in section 2.

1.2 Face Recognition

The dream of tracking individuals by their facial characteristics goes back as far as the 1870s. [20] records how a police records clerk in Paris, Alphonse Bertillon, created a system for taking facial metrics from photographs. Distances between facial features were measured and encoded. Police photographs could be filed according to the codes, which reduced the search space for a suspect — police officers had only to compare suspects with photos which had similar measurements, rather than the entire portfolio.

A hundred years after Bertillon, research began in earnest to automate the recognition of faces by machine. By the end of the 20th century, there had been significant progress. [35] surveys the main algorithms in use, as well as considering the neuroscience and cognitive aspects of human facial recognition, and problems such as variation in pose and illumination.

1.2.1 Face Recognition by Humans

[29] presents 19 observations on the human ability to recognise faces, all of which are relevant to research in computer vision. The results most relevant to the proposed face tracking system are mentioned below.

It is clear that humans adopt a holistic approach to identifying one another. Recognition may be multi-modal (for example, combining face data with body shape or gait), but the studies in [29] indicate that facial information is one of the principal ways in which we recognise each other. A multi-modal recognition system could take metrics such as gait into consideration, combined with facial recognition to increase robustness.

Recognition is not only multi-modal but also multi-method. Humans combine multiple visual cues to recognise a face. In designing machine algorithms for face recognition, we acknowledge that any technique is only part of the solution. A robust recognition system will combine results from multiple techniques.

[29] asserts that the spatial relationship between facial features is more important than the shape of the features themselves. Like humans, machines can accurately recognise faces from a low-resolution image (see [27]) as these spatial relationships are preserved. Skin pigmentation cues are at least as important as shape cues and these are also preserved at low resolution.

Besides the role of pigmentation in recognition, colour is important for segmentation. Especially at low resolutions (where shape information may be degraded), colour information allows humans to estimate the boundaries of image features much more accurately than when presented with a greyscale photo. Likewise, colour cues can help a computer to locate local facial features such as the eyes. Analysis of the hue histogram of an image allows better estimation of the shape and size of the eyes than the luminance histogram alone.

In considering local features, consideration is usually given to the eyes, nose, mouth and possibly the ears. However, the studies in [29] revealed that the eyebrows are one of the most important local features in human facial recognition. Eyebrows

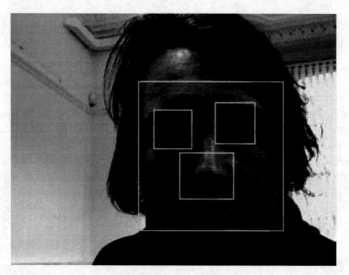

Fig. 1 OPENCV face and local feature detection using nested Haar cascades

have several features which make them suitable for machine recognition: they are stable, with high contrast, which makes them easy to acquire even in a low resolution image. Also, they sit on a convex part of the face, which makes them less susceptible to shadow and illumination problems.

It is interesting to note that faces can be stretched or compressed without any loss of recognisability. Humans may be able to adapt to such transformations because they are similar to the effect of a facial rotation out-of-plane. It is notable that ratios of distances between features in the same dimension remain constant under stretching/compression transformations. Humans may use such *iso-dimension ratios* for recognition. This suggests that machine facial recognition algorithms could adopt the same approach. Instead of the usual approach of pose estimation (see discussion below), the facial metric could be based on ratios which are invariant to changes in scale and rotation. This idea will be developed further in section 4.1.

Finally, we note that humans make temporal associations when observing the movements of someone else's face. This helps to build a more general model of the face.

1.2.2 Face Recognition by Machine

The most important face recognition algorithms and techniques are discussed in [35] and [34].

Holistic, or **appearance-based**, approaches to face recognition were pioneered by Turk *et. al* in their paper on Eigenfaces [30], where differences in human faces are represented by a set of eigenvectors. A covariance matrix is calculated over a set of training images for the face of an individual. The eigenvectors and eigenvalues of the covariance matrix are extracted using Principal Component Analysis (PCA).

This has the effect of projecting the high dimensionality of the face image space into a feature space of lower dimensionality. Features are classified on a nearest-neighbour basis. The technique can be applied to local features as well as whole faces (in this case the Principal Components are referred to as Eigeneyes, Eigennose, *etc.*) Eigenfaces produces very reliable results under laboratory conditions, but is not robust to changes in pose, illumination and expression. It also has high computational requirements, making it unsuitable for real-time facial recognition from a surveillance video stream.

Turk discusses improvements to and extensions of his Eigenfaces technique in [34]. PCA builds its model by minimising the pairwise relationships between pixels in the image training set. The approach can be generalised to minimise the second-order and higher-order dependencies in the image space, using Independent Component Analysis (ICA). ICA attempts to find the basis along which the data are statistically independent. It was first applied to face recognition in [2], where ICA was shown to be more robust than Eigenfaces to minor changes in illumination, expression and appearance (hair, make-up).

Another well-known appearance-based approach is Fisherfaces [3]. The Fisherface approach attempts to improve on Eigenfaces by creating a model which is invariant to changes in illumination and expression (but not pose). Fisherfaces uses Linear Discriminant Analysis (LDA), which is similar to PCA in that the high dimensionality of the image space is reduced to a lower-dimensional feature space. Fisherfaces attempts to choose the direction of the projection such that variations in lighting and facial expression are projected away but the features used for recognition are clustered. PCA chooses the projection which maximises the total scatter, preserving unwanted variations in lighting and facial expression. The computational requirements of the two approaches are similar. Fisherfaces does not solve the problem of variations in pose. The image in the test set must be similar to the image in the training set to be detected. The test set must contain a comprehensive set of images under different lighting conditions.

PCA, LDA and ICA are compared in [28]. Another similar approach which deserves a mention is Tensorfaces [1]. Tensorfaces uses multi-linear analysis, a variation of PCA, to combine training sets of face images under different poses, expressions and lighting.

All the holistic approaches require a sizeable training database of face images under different poses, expressions and lighting to provide accurate results. For a surveillance application, it is unlikely that an accurate model of all possible facial variations can be constructed in advance. The computational requirements are also considerable. These challenges mean that holistic, appearance-based methods have not been demonstrated to work accurately in real-world situations. Turk concludes [34] that Eigenface or other appearance-based approaches must be combined with feature- or shape-based approaches to recognition to achieve robust systems that will work in real-world environments.

Structural, or **feature-based**, approaches to face recognition begin by locating local features such as the eyes, nose and mouth. The locations and characteristics of the features are then used to classify the face. As discussed in [35] and [34], early feature-based approaches were based on pure geometry methods, which match on measurements between features, such as the distance between the eyes or from the eyes to the mouth. A more sophisticated approach is Elastic Bunch Graph-Matching [32], which represents faces using labelled graphs, based on a Gabor wavelet transform. Image graphs of faces are compared with a similarity function. The technique was shown to be robust to small changes in rotation (up to 22°). One advantage of this method is that it requires only a single photo to create the elastic bunch graph.

Some recent developments have attempted to synthesise appearance-based and feature-based methods in a hybrid approach. [36] uses semantic features (the eyes and mouth) to define a triangular facial region. Tensor subspace analysis is applied only to this region to create the feature vector. This has the advantage of combining the geometric information from the spatial locations of the local features and the appearance information from the texture of the facial region. The computational requirements are much less than Eigenfaces or Tensorfaces alone, and the accuracy is higher. However, the problems of variations in illumination, pose and scale are not considered.

1.2.3 Face Recognition from a Video Stream

The machine recognition techniques considered so far were considered in the context of recognising faces from still images. For a surveillance application, we must also consider the challenges (and opportunities) of recognising faces from a video stream. In a surveillance video, we can expect that faces will be captured under variable scale, pose and illumination, and that occlusions will be likely. In our proposed scenario (a Secure Corridor in an airport), we can assume that the environmental lighting is constant, but there will still be variation in terms of lighting direction and shadow.

Almost all of the techniques considered so far assume that it is possible to capture a full-frontal view of the face, but we cannot make this assumption in a surveillance application. Pose angle can significantly decrease recognition accuracy. [13] tackles this problem through a pose-estimation step. Faces are extracted from video frames using an Adaboost classifier, then the same technique is used to locate local features (eyes) within the face area. The location of the centre-point between the eyes relative to the centre-point of the detected face area is used to assign a Gaussian weight to each captured face frame. Face frames which are closer to normal (frontal) pose will be given a higher weight during the recognition step.

The problem of occlusions is considered in [17]. Part of the face under consideration may be occluded, for example by long hair, sunglasses, a scarf or dark shadow. The approach taken is "recognition by parts", where the facial image is broken up into sub-blocks which can be considered individually. Features based on wavelet coefficients are extracted from each sub-block. It is assumed that some of the features are corrupt, but it is not known in advance which ones. A Posterior Union Model is used to find the set of features which maximise the probability of a correct match.

We should also recognise that the video modality gives us certain advantages compared to face recognition from static images. As mentioned in [29], humans observe temporal relationships when recognising someone's face. Our machine facial recognition system can also exploit the temporal relationship between video frames. For example, well-established tracking methods such as particle filtering [11] could be used to track the movement of a face within a video stream. Multiple samples of the same face can then be extracted and passed to the recognition system. As we can take many samples, we can weigh or discard samples which are of poor quality.

Video modality can also be combined with recognition from other modalities using multi-modal fusion algorithms. [13] combines facial recognition with voice recognition to improve accuracy. Gait analysis could also be considered, as it would not require any additional sensors.

1.3 Real-Time Face Recognition from Surveillance Video

We have considered the state-of-the-art in face recognition technology. Face recognition in laboratory conditions is quite mature, but there are many outstanding problems which prevent accurate real-time recognition of faces from a video stream. Our experimental system attempts to investigate some of these problems.

As discussed above, holistic (appearance-based) approaches are not suitable for surveillance applications. They are sensitive to facial expressions and occlusions, and face images must be accurately normalised for pose, illumination and scale. Eigenfaces and similar approaches also have a feature vector of high dimensionality, which makes the process of matching computationally expensive. We have therefore adopted a shape-based method. [9] proposes a shape-based method for recognising faces from a video stream using a combination of a fixed (CCD) and PTZ camera. We have calculated a feature vector following a novel approach suggested by [33].

The recognition system must be robust to changes in illumination, scale, expression, pose and occlusion:

- **Illumination:** detecting local features within a face is reasonably robust to changes in illumination.
- **Scale:** the feature vector is constructed from ratios of lengths and angles, so is invariant to changes in scale.
- **Expression:** we have selected features from the non-deformable parts of the face to provide invariance to changes in expression.
- **Pose:** we do not assume that we will have a frontal view of the subject. Two models are constructed of each subject, one frontal view and one profile view. The main part of the feature vector is constructed from differences in ratios of angles, which provides reasonable robustness to rotation within one view. A pose estimation step is therefore not required.
- **Occlusion:** as we only need to sample features from the non-deformable part of the face, and we extract multiple samples from the video stream, there is reasonable robustness to occlusions.

The feature vector is extracted from the spatial relationships between local features. It is therefore of very small dimensionality, allowing fast transmission over the network and rapid matching to the database.

Our system was developed and tested using the following platforms and test data:

- Face detection was implemented in C using the OPENCV [5] libraries.
- Feature extraction algorithms were developed using MATLAB®. The test data included a live webcam stream, digital photos taken by the authors and the FERET database of face images.
- Feature extraction was then ported to C++/OPENCV. Test data included a live webcam stream and videos of pose variation from CSIT[3].
- The classification and matching system was implemented in C++ using the OPENCV SVM class.

Section 2 describes the design and architecture of our experimental system. Sections 3 and 4 describe the algorithms and operation of the front-end and back-end, respectively. Our experimental results and conclusions are discussed in sections 5 and 6.

2 System Architecture

Our experimental system has been designed with the requirements of a real-time, distributed sensor network in mind. It is envisaged that sensor nodes would be located all around the Secure Corridor. Each node would contain a video camera, an on-board processor and a network connection. The nodes would detect faces, then send data to a back-end for matching to the database. The system architecture is shown in figure 2.

First, faces are detected by the distributed sensor nodes. Feature extraction is also carried out by the node. The high dimensionality of each video frame is reduced to a feature vector of very low dimensionality. This vastly reduces the network bandwidth required and spreads the processing load across the distributed network. Object tracking could be carried out by the nodes, so that multiple face samples are attributed to the same person. Feature vectors can then be tagged with an identifier and sent to the back-end. (Object tracking was not implemented in our experimental system).

The back-end system receives feature vector packets and unpacks them for pattern matching with the database. This is treated as a classification problem, with each individual enrolled on the system representing a class. Matching is performed using a Support Vector Machine (SVM). Temporal relationships between frames could be exploited to increase the likelihood of a correct match. In a production systen, the back-end would log the time and location (camera no.) of individuals under surveillance, and would be configured to send alarms to operators.

---

[3] Centre for Secure Information Technologies, Queen's University, Belfast (http://www.csit.qub.ac.uk/). Thanks to Dr. Darryl Stewart and Adrian Pass for providing the video database [24].

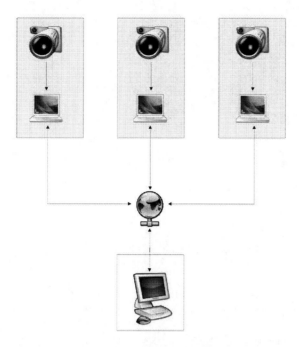

Fig. 2 Distributed System Architecture: Multiple Front-Ends

Please see the Appendix to this chapter for more implementation details, including how to obtain the source code.

3 Front-End

The front-end has three functions: to detect faces, to track detected faces and to extract features for transmission to the back-end recognition system. The process is summarised in figure 3.

3.1 Face Detection

Face Detection is carried out using the OPENCV Face Detector [5]. This is an implementation of the Viola-Jones Face Detector [31], which uses a boosted rejection cascade based on Adaboost [8].

The theory of boosting is that many weak classifiers can be combined to make a strong classifier. Adaboost is a supervised classifier, *i.e.*, it takes a set of labelled samples as input. A tree of weak classifiers is created by applying a weak algorithm to the labelled data. Each weak classifier is assigned a weight based on its performance on the training data. Classifiers that return the wrong result are given a higher weight, to concentrate attention on the places where errors are likely to be

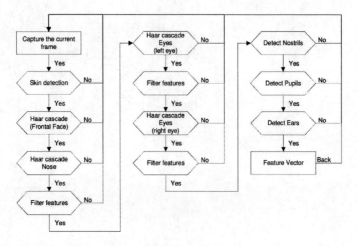

Fig. 3 Front End: Face Detection and Feature Extraction

made. If this process is applied iteratively, the training error of the strong classifier exponentially approaches zero. This is repeated until the error rate drops below a set threshold.

Viola-Jones uses Adaboost both for feature selection and to train the face detector. During feature selection, Adaboost can select a few important features from the set of all possible features. The features are Haar-like wavelets. The Haar-like features are extracted by applying a threshold to the sums and differences of rectangular image regions. OPENCV also includes diagonal wavelets[4] to detect rectangular features rotated 45°. Each weak classifier in OPENCV typically uses one feature (at most three).

A weak classifier $h(x, f, p, \theta)$ is used, where x is a 24×24 pixel sub-window of an image, f is a feature, θ is a threshold and p is the polarity (direction) of the inequality:

$$h(x,f,p,\theta) = \begin{cases} 1 \text{ if } pf(x) < p\theta \\ 0 \text{ otherwise} \end{cases}$$

Each stage of the boosting process attempts to find the Haar-like feature which best separates faces from non-faces.

One of Viola-Jones' innovations was the use of integral images to rapidly compute the Haar-like features at any scale or location in the image. The integral image at (x, y) is defined as the sum of all pixels above and to the left of (x, y). The sum of all pixels within any given rectangle can be rapidly calculated from the sums and differences of the four corner pixels of the integral image. By thresholding these rectangles, the Haar-like features can be calculated at any location or scale without needing to build a multi-scale pyramid.

---

[4] This extension to the Viola-Jones technique is described in [16].

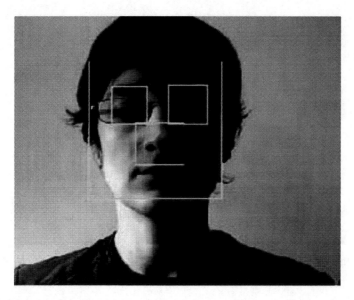

Fig. 4 OPENCV face and local feature detection

Another important innovation was the use of a cascade of classifiers to reduce the computation time and increase accuracy. First, a two-feature strong classifier is applied to the integral image. The threshold is set to minimise false negatives at the cost of a high false positive rate (up to 50%). This rapidly rejects background regions and focuses attention on more promising regions. Regions which pass the first classifier are passed to the next classifier in the cascade. The strategy is to reject negatives as early as possible and gradually reduce the number of false positives as the regions pass through the cascade. A region that passes through every level of the cascade is classified as a face.

The Viola-Jones method can be applied to other objects besides faces, but works best on objects with "blocky" features rather than those where the outline is the most distinguishing characteristic. This is because the classifier must include part of the background of an object in its model of the object's outline.

OPENCV includes pre-trained cascades for frontal faces, profile faces, eyes, noses and mouths. Our implementation used nested cascades within the OPENCV face detector to find local features (eyes and noses) within detected faces (see figure 4). Feature extraction is discussed in section 3.3.

The detection of profile views works well when the background is plain but less well with a textured background, for the reason mentioned above. The detection of profile faces was improved using skin detection (see below).

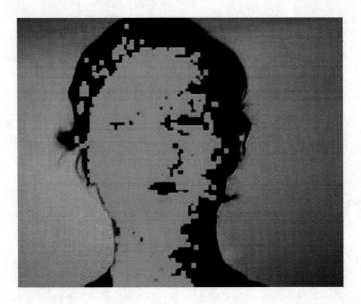

Fig. 5 Skin Detection

3.1.1 Skin Detection

Viola-Jones operates on greyscale images, but for our implementation, we decided to also exploit the colour information in the video stream[5]. By applying skin colour detection to the input video frames, it is possible to divide the search space into face candidate regions and background regions. Face detection need only be applied to face candidate regions. The segmentation of skin regions based on luminance and chrominance values is discussed in [10].

OPENCV's Adaptive Skin Detector was used to detect face candidates (see figure 5). Skin-coloured areas (with a margin added around the edges to allow for areas of shadow) were passed to the face detector. This resulted in a significant speedup and also improved accuracy, by discarding face-like regions which are not skin-coloured (false positives).

Another benefit of using the skin detector was to segment faces in profile view from the background. This enabled more accurate detection of the facial profile, overcoming the problem of detecting an outline against a textured background (see above).

The results of skin detection are discussed in detail in section 5.1.

---

[5] Indeed, the original Viola-Jones paper [31] suggests that alternative sources of information can be integrated with the basic approach, to achieve even faster processing and better frame rates.

3.2 Face Tracking

As we are examining video sequences rather than still images, we can exploit the temporal relationships between frames to improve recognition accuracy. It is possible to track the movement of individual faces in a video stream using techniques such as Boosted Particle Filtering, Mean Shift, or Kalman Filtering[6].

If we can successfully track a moving face through a video sequence, we can take each frame as being a sample of the same person. Extracted features can be tagged with an identifier before sending to the back-end for recognition. Features with the same identifier can be considered as multi-samples of the same individual.

Note: Multi-sampling was not implemented in the experimental system described in this chapter.

3.3 Feature Extraction

One of the objectives of the face recognition system is invariance to changes in pose. In previous approaches, this problem has been addressed by either creating a 3D model [1] or by employing a pose estimation step [13, 32]. The creation of a 3D model is problematic, as it requires a large set of training images for each individual. This is impractical in a real-time surveillance application. Our system follows a novel method, proposed in [33], where the feature vector is invariant to pose, so a pose estimation step is not required.

[31] notes that the face detector can detect faces which are tilted up to about $\pm 15°$ in plane and $\pm 45°$ out-of-plane (towards a profile view). In order to recognise faces from any angle, our system constructs two separate models of each individual: a frontal model and a profile model.

Each model (frontal and profile) defines a **baseline**, a **reference length** and a set of **feature key points**. The feature vector is defined by the spatial relationships between these features (see section 4.1).

3.3.1 Frontal Face Model

For the front face model, we first define two reference points — the centre points of the irises. The front baseline is defined as a line connecting the reference points (see figure 6. This length of this line is used as the reference length for the front face. Next, we define a number of key points which denote the location of shape-based features such as the nostrils and the tips of the ears. The feature vector is calculated from a set of lengths and angles measured from the baseline to the feature key points.

The outline algorithm for finding the baseline and key feature points is as follows (see figure 3):

---

[6] Linear dynamic models are discussed in [7]. Non-linear models are also discussed in the unpublished chapter, "Tracking with non-linear dynamic models", available at http://luthuli.cs.uiuc.edu/ daf/book/bookpages/ pdf/particles.pdf. See also [11].

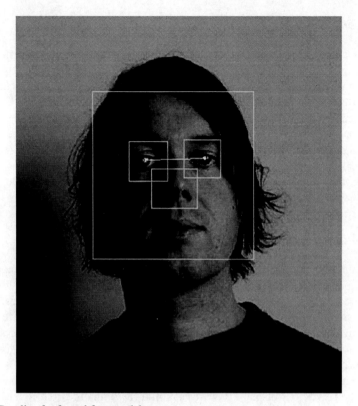

Fig. 6 Baseline for frontal face model

1. Find the face bounding box using the OPENCV Haar cascade detector
2. Use nested Haar cascades to detect local features (eyes and nose)
3. If local features are not detected, discard the frame
4. Search within eye bounding boxes for reference points
5. Search within nose bounding box for feature key points
6. Search for other key points (for example around face contours)

Note that the reference points and feature key points are chosen from features in the non-deformable part of the face, to provide robustness to changes in facial expression. The detailed algorithms for extracting the reference points and other key points are described in the following sections.

3.3.2 Iris Detection

The reference points for the front face are the centres of the two irises. During preprocessing, the eyes have been located and the system has determined the bounding box for two eye regions. Face candidates which do not have two eye regions have been discarded. If more than two eyes have been detected, the system attempts to estimate the most likely eyes by considering their location and size relative to each other and the face bounding box.

The reference points are located at the centre of the iris/pupil within the two eye bounding boxes. There are a number of possible approaches to iris detection. If sufficient training data were available, it would be possible to add a third nested Haar cascade to the OPENCV face detector to detect irises within the eye region. [6] proposes an integrodifferential operator which can detect iris and pupil boundaries. Another possibility would be to match curves to the eye region using a Hough Transform. Circles of approximately the right size could be extracted as irises and pupils. [14] takes this approach, improving robustness by using a separability filter.

Our approach was a geometric approach using thresholding and blob detection on the image. This has the advantage of being simple and fast to compute.

Detection uses the following algorithm[7]:

1. Use a 5×5 averaging filter to remove noise and improve blob detection[8].
2. As the surface of the eye is highly reflective, there is usually a point of light on or near the pupil/iris. To locate this point, a gradient is calculated over the image. Small gradients are discarded. Large gradients are weighted by the distance from the centre of the bounding box. The highest weighted gradient is taken as a point assumed to be on or close to the pupil/iris (see figure 7).
3. Histogram equalise to improve blob detection.
4. Threshold image at $0.25 \times \max(image)$.
5. Erode thresholded image with a 3×3 structuring element (see figure 8)
6. Find the blob closest to the gradient point.
7. Use region selection to find the maximum and minimum points within the iris blob.
8. Calculate the centre of the blob as the average of the maxima/minima. (figure 9)
9. **return** centre of blob

Note that this method works reasonably well on people who are wearing glasses. [3] solves the problem of wearing glasses by dividing images of an individual into two classes, one which is "wearing glasses" and one "not wearing glasses". The OPENCV face detector includes eyes wearing glasses in its cascades for face and eye detection.

Once the two reference points have been determined, the baseline is calculated as the line joining the two points (see figure 6).

3.3.3 Nostril Detection

Having detected the two reference points, the nostrils are detected next, as the first feature key points. The nostrils are important, as they are rarely occluded and

---

[7] Detection algorithms were developed experimentally in MATLAB (see detect_iris.m) and subsequently implemented in C++/OPENCV.
[8] Later experimentation with a larger test set showed that the averaging filter did not improve detection accuracy, so this step was removed from the C++ implementation.

Fig. 7 Eye region: blurring filter and location of highest weighted gradient

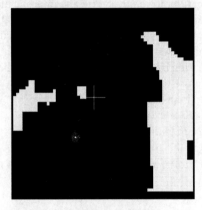

Fig. 8 Eye region: thresholding and blob detection

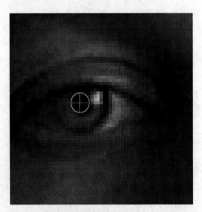

Fig. 9 Eye region: detection of centre of pupil/iris

establish the location of the centre of the face. The location of the nostrils is used to calculate a centre-of-face reference line which is perpendicular to the baseline.

Nostril detection is similar to iris detection. It would be possible to use a third nested Haar cascade if we had a training set of nostril shapes. Again, it would be possible to use a Hough transform to detect the roughly-elliptical nostril shapes. We adopted a geometric approach similar to the method used for iris detection. This works very well, as the nostrils appear clearly as two dark blobs in the thresholded image. The detailed algorithm is as follows:

1. Use a 5×5 averaging filter to remove noise and improve blob detection[9].
2. A gradient is calculated over the image and used to estimate the direction of illumination. Typically, illumination is from one side, meaning that one side of the nose is illuminated and the other side is in shadow. The highest gradient on the blurred image is shown in figure 10.
3. Local histogram equalisation is performed independently on both sides of the nose (see figure 11). This compensates for unequal lighting.
4. Threshold image at $0.25 \times \max(image)$.
5. Erode thresholded image with a 3×3 structuring element (see figure 12)
6. Find the two blobs closest to the estimated tip of the nose[10].
7. Use region selection to find the maximum and minimum points within the nostril blobs (see figure 13).
8. **return** nostril boundary points

Once the nostril points have been detected, it is possible to calculate the centre face reference line. A point is calculated as the midpoint of the two nostrils, and a line perpendicular to the baseline is drawn through this point (see figure 14). The point where the baseline and centre face line intersect is used as the reference point for key feature measurements.

3.3.4 Ear-Tip Detection

The third set of features that we use for the facial feature vector are ear-tips. Unlike irises and nostrils, which can be detected as blobs on the front of the face, ear-tips can be detected as a corner along the edge of the face.

If a suitable set of training data was available, it would be possible to detect ears using a nested Haar cascade as we do for eyes and nose. However, this introduces the same problems as profile detection, in that it is difficult to train a Haar cascade to detect edges reliably in the presence of textured backgrounds. This could be compensated for using skin detection as previously discussed.

As we have already located the eyes, we use this as a cue to detect the location of the ears. As the Haar cascade face detector sometimes returns a bounding box which has cropped the ears, we expand the detection area to compensate for this.

---

[9] Later experimentation with a larger test set showed that the averaging filter did not improve detection accuracy, so this step was removed from the C++ implementation.

[10] Initially our algorithm was thrown off by reflected light from Adrian's nose-ring but we were able to compensate for this!

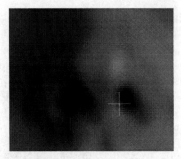

Fig. 10 Nose region: blurring filter and location of highest weighted gradient

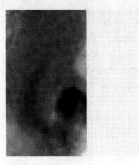

Fig. 11 Nose region: local histogram equalisation

Fig. 12 Nose region: thresholding and blob detection

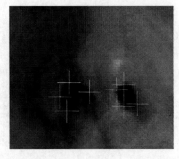

Fig. 13 Nose region: detection of boundary points

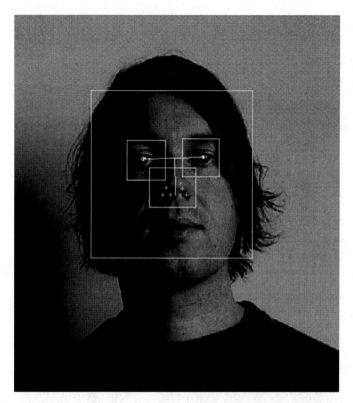

Fig. 14 Front Face Model, showing Baseline, nostril feature key points and centre reference line

The algorithm that we used for ear-tip detection is as follows:

1. Detect within the region returned from the OPENCV face detector.
2. If the skin patch representing this face is wider than the bounding box returned by the Haar cascade, expand the detection region as required.
3. Use OPENCV morphological operations to smooth the image and remove internal edges. This leaves the outside contour of the face as the only strong edge.
4. Perform Canny edge detection to get the face outline (see figures 16 and 17).
5. Detect corners by calculating gradients along the face edges
6. Choose the lowest sharp corner on either lateral side of the face. This should be located where the ear lobe meets the face outline.
7. **return** ear tip points

Ear detection is much less reliable than detection of local features in the front-of-face, as we cannot guarantee that the ear will not be occluded. Even if the ears are clearly visible, our algorithm may fail if the jaw-line is within the region of interest, as this will result in another sharp corner lower than the ear.

Detected ear-tips are shown in figure 15.

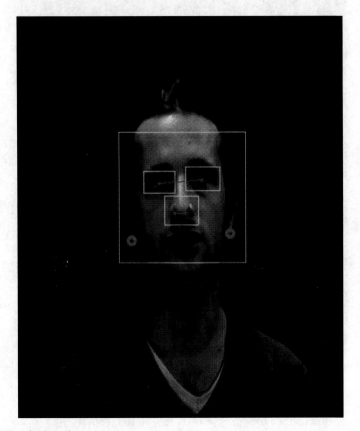

Fig. 15 Ear-tip detection

3.3.5 Profile Face Model

The model for the profile view is constructed using the same principles as the frontal face. The image is searched using the OPENCV face detector with the cascade for detecting profile faces. When a face is detected, a nested Haar cascade is used to detect an eye (naturally, only one eye is visible in profile view).

The baseline for profile view is again constructed using two reference points. The first point is at the top of the nose. The second point is located by fitting a line to the profile of the nose, and then finding a perpendicular line which is tangential to the bottom of the nose. The point where this line meets the upper lip is the second reference point. The baseline is constructed by joining these two points. The length of the baseline is the reference length for the profile view. The profile baseline is shown in figure 18.

Real-Time Face Recognition from Surveillance Video 175

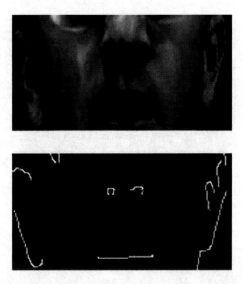

Fig. 16 Ear outline detection: Darryl

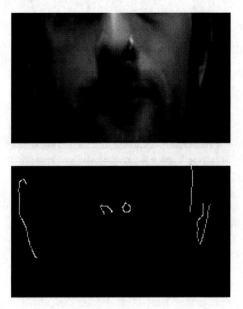

Fig. 17 Ear outline detection: Adrian

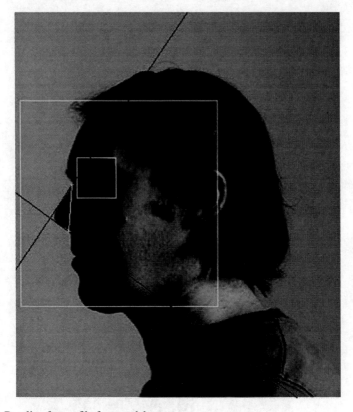

Fig. 18 Baseline for profile face model

OPENCV provides a Haar cascade only for left-facing profiles. Normalising all training samples to the same view is a more efficient use of the data. In order to detect right-facing profiles, the images must first be horizontally flipped.

As previously discussed, using the OPENCV face detector in profile view is not as reliable as frontal view, because the classifier relies on block features. It is interesting to note that the ear in frontal view is an outline feature, but in profile view it becomes a block feature. It would therefore make sense to use the ear as a key feature in profile view. Ear detection using cascaded Adaboost from a profile view is discussed in [12].

If the OPENCV face detector is trained with profile faces, the classifier has to attempt to learn the background variability beside the profile edge. The OPENCV Haar cascade provided for profile faces works reasonably well against a plain background, but to provide reliable detection against variable backgrounds, it would need to be trained with a much larger data set with highly random backgrounds [5].

Viola-Jones requires a very large set of training data. There would need to be thousands of positive examples of faces, and tens of thousands of non-faces. When training a cascade, there should be many more negative examples than positive examples, so that there are enough false positives detected by the early stages of the cascade that can be passed on to train the later stages of the cascade. The data should be well-separated (for example, not mixing training sets in different poses) and well-segmented (box boundaries should be consistent across the training set). These considerations mean that retraining the classifier is a formidable problem.

Note: Feature extraction from a profile image was demonstrated in principle using MATLAB code, but due to the issues discussed in this section, we did not develop a full implementation in OPENCV.

4 Back-End

The feature set extracted for each face is very small — simply an (x,y) coordinate for each reference point and feature key point. These points are labelled and sent to the back-end for analysis (inside a TCP/IP packet). Labels could include camera number, time-stamp, frame number, face bounding box within frame and an object tracking ID (if face tracking using particle filtering has been implemented).

In the back-end, a feature vector is calculated from the set of feature points, and this feature vector is compared to the database of enrolled people using a SVM (Support Vector Machine).

4.1 Feature Vector

To ensure robustness against changes in scale and rotation, only ratios of lengths and angles are stored in the feature vector. First, the distances between the reference points and the other key points are calculated, and expressed as a ratio to the reference length. As all distances are expressed as ratios, they are invariant to changes in scale.

Second, the angles between the reference points and other key points are calculated. The orientation of the baseline is used as the horizontal axis for the coordinate system of the feature vector. *i.e.*, the coordinates of the key points must undergo a matrix transformation into the new coordinate system. Normalising the key points to the baseline means that the feature vector is invariant to rotations in-plane.

The calculated angles do not themselves form part of the feature vector. Rather, the feature vector stores the difference between tangents of the angles formed at each reference point. Taking the difference between tangents means that the feature vector is invariant to rotations out-of-plane. A proof of this result can be found in [33].

The feature vector for the profile view is calculated in the same manner as the frontal view. The key point coordinates are transformed into a coordinate system using the profile baseline as the horizontal axis. The lengths and angles between the reference points and the other key points are calculated. These are encoded in the feature vector as ratios of the profile reference length and difference between tangents of angles. As with the frontal model, the feature vector is robust to changes in scale and pose (rotation).

4.2 Feature Vector and Matching

The set of reference points and feature key points is transmitted to the back-end over TCP/IP. The back-end calculates the feature vector from this set of points following this algorithm:

1. Calculate the gradient of the baseline (the line connecting the irises)
2. Use the gradient to calculate a rotation matrix to normalise the coordinate space, so that the baseline is parallel to the horizontal axis. This ensures that the feature vector is invariant to rotations in-plane
3. Rotate all the points about the left iris
4. Find the face centre line by taking the midpoint between the nostrils
5. Find the intersection of the face centre line and the baseline
6. Normalise the coordinate space so that the intersection of the baseline and face centre line is $(0,0)$. A graphical representation of the normalised feature points is shown in figure 19.
7. Calculate a set of lengths connecting the intersection point and the irises to the key feature points. These lengths are expressed as ratios of the baseline. This ensures invariance to scale
8. Calculate a set of tangents of angles between the intersection point and the key feature points. The feature vector stores only differences between these tangents. This ensures invariance to rotations out-of-plane[11]

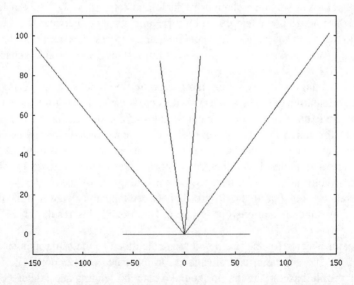

Fig. 19 Normalised feature points for four feature key points (two nostrils and two ear tips). The baseline is the horizontal line at the bottom and represents the line that joins the irises.

---

[11] See [33] for proof.

A set of feature vectors was calculated over a subset of the test videos and used to train the OPENCV SVM. The results of training were saved as an XML file that was used for verification and testing (see section 5.4).

5 Results

The aim of our experiments was to detect faces in a video stream, and extract a reliable set of features as rapidly as possible. The extracted features were then tested for discrimination between different subjects using a Support Vector Machine (SVM).

The process of face detection and local feature extraction is summarised in figure 3. It can be seen that a candidate face image can be rejected by the system at a number of points. In order to be passed to the back-end, the face candidate must pass skin detection, the nested Haar cascades for global and local features and a suitable set of local features must be identified and extracted. As our highest considerations were speed and reliability, we simply dropped any frames that could not pass all these tests. During the experiments, we attempted to reduce the number of dropped frames without compromising reliability.

Our feature extraction algorithms were developed and tested using still images from the FERET database. In order to test how our programs worked on video streams, we used a database of videos of speakers in different poses from CSIT (see section 1.3).

5.1 Skin Detection

The OPENCV adaptive skin detector was used to filter the image and to pass face candidate regions (rather than the whole image) to the Haar cascade detector. Our results show that performing skin detection on the video frames before passing them to the Haar cascade improved both speed and reliability.

Use of skin detection helped to eliminated false positives (non-face areas that are detected as faces). This could also include faces which are not real human faces (*e.g.*, a black-and-white photo of a face on a poster or T-shirt).

Sometimes the OPENCV Haar cascade detector finds the same face object at different scales (see figure 20). By searching skin regions, we can narrow the search to find the most meaningful match.

We experimented with using windows of different sizes around detected skin patches. The results are shown in figures 21 and 22. It can be seen that constraining the search area too tightly reduced the accuracy of the results. This is because areas of shadow to the sides of the face are sometimes not detected as skin. The Haar cascade detector therefore needs an area wider than the skin patch for a successful detection. As we increased the search window around the skin patches to include these shadowed areas, the accuracy of the Haar detection gradually increased until (at around 30–50 pixels) it was as accurate as searching on the whole image.

Obviously the greatest speedup is achieved with a smaller search window. As we increased the size of the search window, the speedup decreased exponentially.

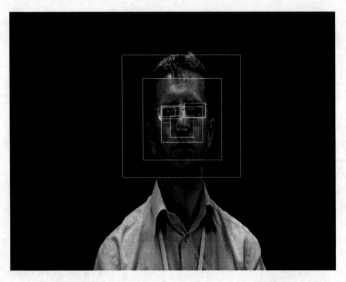

Fig. 20 Face matched at multiple scales using OPENCV Haar Cascade face detector

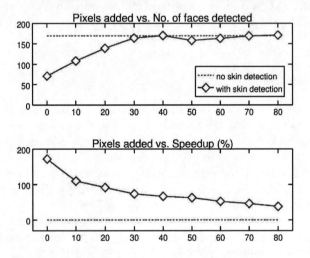

Fig. 21 Affect of Skin Detection on Performance: Darryl

However, a window of 30–50 pixels is still measurably faster than not using skin detection.

Skin detection is also vulnerable to false positives and this reduces performance. This is illustrated in figure 22, as the colour of Adrian's clothing is within the range of detected skin tones (see figure 27; the outer bounding box indicates the detected skin region that is passed to the Haar cascade for face detection). In this case, there is no performance improvement by using skin region detection.

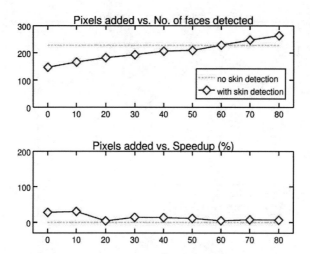

Fig. 22 Affect of Skin Detection on Performance: Adrian

The strategy of skin detection would not be of any benefit if large areas of the background were skin-coloured. In a real system, whether or not to use skin detection could be configured at each sensor.

For our experimental system, we chose a window of 40 pixels around skin regions as giving the best compromise between performance and accuracy.

5.2 Face Detection

The most important features for the OPENCV face detector are the eyes and nose. During informal experiments, it was observed that occluding the eyes or nose prevents a face from being detected. The mouth is a less important feature. It is often possible to occlude the mouth without affecting face detection. In any case our system does not extract features from the mouth area as we restricted our system to detecting features in the non-deformable part of the face.

As illustrated in figure 3, if we did not detect a suitable face and local features within a frame, that frame would be discarded. Figures 23 and 24 show how many face candidates are rejected at each stage of the detection process.

Note that the OPENCV face detector returns many false positives. In our tests, each frame of video contained exactly one face, but figure 23 shows that the number of faces returned by the Haar cascade is greater than the number of frames. The results show that skin detection slightly reduced the number of false positives (while speeding up the overall detection rate).

More face candidates are rejected by the nested Haar cascades which detect local features (eyes and nose). If the system cannot detect two eyes and a nose, the face candidate is discarded. Finally, we identify reference points and feature key points

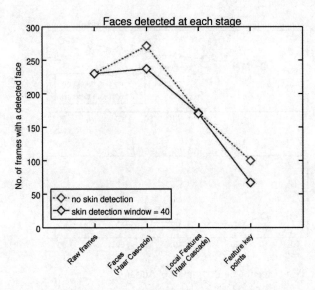

Fig. 23 Rejection of unsuitable face candidates at each stage of the detection process: Darryl

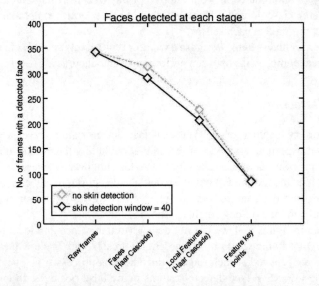

Fig. 24 Rejection of unsuitable face candidates at each stage of the detection process: Adrian

within the local features. If these cannot be accurately identified, the face candidate is discarded. Recognition is attempted only on face candidates which pass all stages of the detection process. Thus, we are discarding around 60%–70% of face candidates in order to improve recognition accuracy.

5.3 Feature Extraction

5.3.1 Morphological Operations

In order to optimise blob detection, we experimented using a number of convolution kernels:

- Custom kernels for erosion and dilation
- Built-in convolution kernels for OPENCV
- Morphological transformations for opening and closing the image

The morphological operations of *closing* and *opening* an image predefined in OPENCV represent a combination of erosion and dilation. The effects are as follows:

Opening an image results in removing small bright regions of the image. The larger bright regions are isolated but retain their size. We tried opening the image because a relatively small bright region (appearing as an artefact of wearing a nose ring) was interfering with our algorithm of nostril detection.

Closing an image results in joining bright regions within the image. The dark regions remain dark and their size is unchanged. We tried this, aiming to provide a better contrast between the nostril/pupils and the brighter areas that surround them.

Opening then closing an image should theoretically discard the smaller brighter regions and then join all the larger brighter regions while keeping the dark regions almost unchanged. Although this makes the nostrils more obvious to the human eye, OPENCV blob detection (based on contour detection) fails because this operation has the effect of destroying edges.

The test video images were convolved with 12 different kernels, and the performance of each was compared. Figure 25 illustrates how the choice of convolution kernel affects the accuracy of feature detection. Each kernel was also compared for its effect on computation time, as shown in figure 26[12].

Our experiments showed that convolving the images with a rectangular kernel of size 2×2 provided the most accurate detection of key features (pupils and nostrils). There was a slightly higher cost in terms of processing time, but this was deemed to be an acceptable trade-off.

5.3.2 Filtering Local Features

Local features were detected using a nested Haar cascade. Often the Haar cascade would return one or more false positives for each local feature as well as the correct location of the feature. It was therefore necessary to filter the local features to determine which candidate was the correct one.

---

[12] The time in seconds refers to the time to process the entire video sequence. The video sequences were 342 and 230 frames respectively, so our system is somewhat slower than real time. Real time speeds could easily be achieved in a production system by exploiting hardware acceleration or parallel processing.

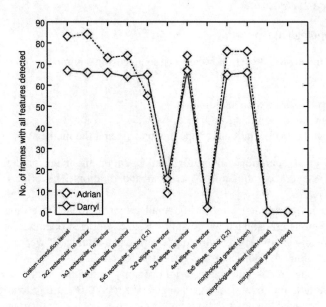

Fig. 25 Analysis of choice of erosion/dilation structures on detection accuracy

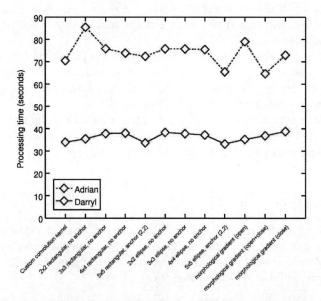

Fig. 26 Analysis of choice of erosion/dilation structures on performance

Several general rules were set when filtering the features:

- The right eye should be in upper right part of the face
- The left eye should be in the upper left part of the face
- The nose should be in the central region of the face. Sometimes the bounding boxes for the nose and the bounding boxes for the eyes can briefly overlap.

The proportions of the human face have been studied for many years, in particular by surgeons specialised in reconstructive and plastic surgery. Thus, medicine has determined that the height of the nose relative to the frontal face should be around 47% [23]. The frontal face Haar cascade often crops the lower part of the face (see figures 27 and 28). This cropping is very unpredictable, so the face height is unreliable for filtering features. However, the nose is never cropped. Since the nose height is not affected by cropping, we tried to use it in defining a ratio that would give the approximate position of the top part of the eyes.

The human frontal face can be divided into three horizontal regions. The aesthetic ideal is that the three regions are the same height, but in our system we allow some flexibility. Since we are interested in determining a relation that would give us the position of the eyes relative to the top of the face, we are only interested in the upper region. This lies between the trichion[13] and glabella[14]. This is approximately 33% of the entire facial height but since the Haar detection of the frontal face crops this part, we calculate this region as being between 20–30% of the face height. The correct position of the eyes should be detected just under the glabella. So the position of the eyes on y axis should be approximately 0.2–$0.3 \times face\ height$ (1) and the height of the nose should be approximately $0.47 \times face\ height$ (2). By substituting $face\ height$ from (2) into (1) it follows that the eyes are expected at approximately 0.42–$0.63 \times nose\ height$.

We conducted experiments to determine the coefficient that works best within this interval. Table 1 shows the number of faces (containing two eyes and a nose) found in the video stream for different values of the coefficient.

The value 0.53 was chosen for the facial proportion coefficient, because it provided good detection rates and no false positives. 0.5 provides even better detection rates but was considered too close to the value that outputs false positives.

In our test system, the position of the top of the nose bounding box was approximated using a fixed number of pixels. The detection was calibrated for the size of the faces in the test videos. In a real system, this parameter would have to be automatically calibrated to work on a wider range of images.

The proportions that were used to choose the best nose and best eyes from all candidates are shown in figure 29.

---

[13] The point where the hairline meets the midpoint of the forehead.

[14] Reference point in anthropology representing a smooth elevation in the frontal bone just above the bridge of the nose.

Fig. 27 Face detected using Haar Cascade, but upper and lower regions are cropped

Fig. 28 Face detected using Haar Cascade, but upper and lower regions and ears are cropped

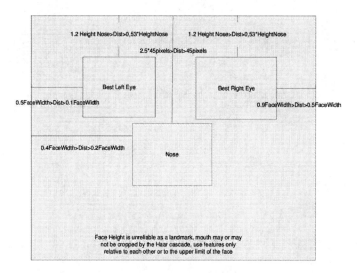

Fig. 29 Feature Extraction: Selection of best local features from within the face bounding box

Table 1 Effect of coefficient for facial proportions on detection accuracy. The Adrian video had 342 frames and the Darryl video had 240 frames.

Subject	Coefficient	No. of faces detected
Adrian	0.48	316[15]
	0.5	280
	0.53	205
	0.55	128
	0.6	21
Darryl	0.48	170
	0.5	170
	0.53	170
	0.55	168
	0.6	117

5.4 Feature Vector and Recognition System

Face recognition in the back-end was implemented using the OPENCV SVM class. A set of training videos was processed by the front-end. Feature points from the training set were sent to the back-end, which generated feature vectors and dumped

---

[15] 316 detections includes some false positives — in some cases, eyebrows are detected as eyes.

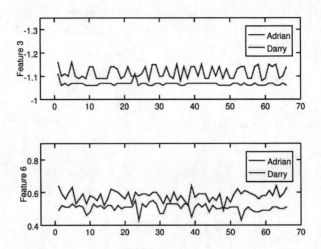

Fig. 30 Discriminative capacity of the feature vector: comparison of two features for two subjects

them into a file. A matrix was initialised with each feature vector as a row and we specified how many rows represented training data for each subject. The result of training was saved in an XML database.

The SVM was trained for two subjects using one video for each subject. In the training videos, the subjects were looking straight ahead. The video for Adrian was 13 seconds long and 67 faces were extracted and passed to the SVM for training. The video for Darryl was 9 seconds long, and 83 faces were extracted. The number of frames extracted for Adrian is lower because in many frames his eyes are closed, so those frames are discarded. Also, Adrian's nose ring made accurate feature extraction more challenging for the front-end! The feature vector in our experimental system is of low dimensionality, with only six features. Figure 30 illustrates the discriminative capacity of our system, showing a comparison of features 3 and 6 for the two subjects. Features 1, 2, 4 and 5 measured ratios of distances between features (relative to the reference length), whereas features 3 and 6 measured the differences of tangents between the reference points and feature key points. Features 3 and 6 were therefore the most robust to rotations out-of-plane and provided the greatest discrimination between the subjects in our experiments.

We demonstrated that it is possible to calculate more lengths and tangents between the feature points we have already extracted. A production system would add additional feature points to provide a feature vector of larger dimensionality and therefore capable of greater discrimination.

Testing was conducted using a different video from the training set for each subject. The video for Darryl was 9 seconds long and he was rotated 10% from frontal view. The video for Adrian was 11 seconds long and was facing straight ahead. The videos were processed by the front-end and feature points were sent to the back-end for identification. The back-end calculated the feature vector for each frame and returned identification results as shown in table 2.

Table 2 Results of identification test for two subjects

Subject	Faces found (front end)	Faces identified (back end)	False positive	False negative
Adrian	20	20	0	0
Darryl	38	38	0	0

These tests can be considered as a proof-of-concept. It would require a larger set of test data to do an accurate comparison of the discriminative capacity of the feature vector.

It was not possible to test on rotations larger than 10% as we were using the ear tips as a feature, and at greater than 10% rotation, one ear is occluded. This could be compensated for by adding more features and by using missing feature theory to ignore the effects of occluded features.

6 Conclusions

This project has demonstrated a system for face recognition from surveillance video. The system was demonstrated to be robust to changes in illumination, scale, facial expression and reasonably robust to occlusions and changes in pose. Attention was given to performance considerations, and the system can operate in real time. All of these features make this approach suitable for a video surveillance application.

One of the notable findings was the improvement in both performance and accuracy of the OPENCV face detector, when the Viola-Jones face detection was combined with an analysis of colour skin-tone information. Colour information could be exploited further to improve the detection of local features as discussed in [10]. Our feature-detection algorithms were demonstrated to be invariant to changes in lighting direction.

The local features that we used were irises, nostrils and ear-tips. Irises and nostrils are reasonably straightforward to extract in frontal view, although reflected light from glasses or a nose-ring can degrade performance. Ears are more difficult to locate as Viola-Jones' method is better at detecting "blocky" features rather than outline features. Ears can also be occluded by hair, earrings or self-occluded by a rotation of the face. Nonetheless, it was observed that the difference in tangent angles to the ear-tips was one of the most discriminant features in the feature vector.

In profile view, the key features were the location of the iris and the shape of the nose profile. The iris may not be available if the person is wearing glasses, as the stem of the glasses usually occludes the eye. It is easy to detect the ear in profile view (if it is not occluded) as it becomes a "blocky" feature when viewed from the side.

In a real-world system, it would be necessary to select more features than we have outlined here to provide sufficient discrimination between people enrolled on the system. Suitable features could include corners and edges on facial features such as the eyebrows and mouth. We have already mentioned that the eyebrows are one of the most important features in human facial recognition [29]. Features in the non-deformable part of the face should be accorded more weight, to provide robustness to changes of expression. We have also mentioned that the full set of features may not be available in every case. It may be necessary to select a different set of features for different subjects. Missing-feature theory could be applied to this problem [17].

It is unknown how many features are required to provide discrimination over a larger population. This could be the subject of a future study. We envisage that this type of system would be used in a controlled environment (such as a Secure Corridor in an airport) where the population is of limited size and people can be enrolled onto the system as they enter the Corridor.

It is likely that the best results will be achieved by combining this approach with other methods. Colour information was used in detection but not as part of the recognition system. [29] states that pigmentation is an important recognition cue for humans (at least as important as shape), so it would be natural to include skin pigmentation, hair colour or eye colour as part of the feature vector.

Our shape-based approach could also be combined with appearance-based approaches in a hybrid detection system. Just like Bertillon in the 19th century, local feature recognition could be used to narrow the search space and then global features could be used to get an exact match. It is also possible to use appearance-based methods on local features (Eigeneyes, *etc.*). In this case, our method could be used to perform pose estimation. [19] discusses how to improve face recognition using a combination of global and local features.

The accuracy of facial recognition can also be improved by combining face detection with other forms of detection using multi-modal fusion. Gait analysis would be a likely candidate as it does not require any additional sensors.

Finally, it is worth mentioning that this approach to machine vision also has non-security applications, for example the recognition of human faces by a robot or computer game.

References

1. Alex, M., Vasilescu, O., Terzopoulos, D.: Multilinear analysis of image ensembles: TensorFaces. In: Heyden, A., Sparr, G., Nielsen, M., Johansen, P. (eds.) ECCV 2002. LNCS, vol. 2350, pp. 447–460. Springer, Heidelberg (2002)
2. Bartlett, M., Movellan, J., Sejnowski, T.: Face recognition by Independent Component Analysis. IEEE Transactions on Neural Networks 13(6), 1450–1464 (2002), doi:10.1109/TNN.2002.804287
3. Belhumeur, P., Hespanha, J., Kriegman, D.: Eigenfaces vs. Fisherfaces: Recognition using class specific linear projection. IEEE Transactions on Pattern Analysis and Machine Intelligence 19(7), 711–720 (1997)

4. Bentham, J.: Panopticon; or, the inspection-house: Containing the idea of a new principle of construction applicable to any sort of establishment, in which persons of any description are to be kept under inspection; and in particular to penitentiary-houses (1843), http://oll.libertyfund.org/
5. Bradski, G.R., Kaehler, A.: Learning OpenCV (2008)
6. Daugman, J.: High confidence visual recognition of persons by a test of statistical independence. IEEE Transactions on Pattern Analysis and Machine Intelligence 15(11), 1148–1161 (1993)
7. Forsyth, D., Ponce, J.: Computer vision: a modern approach. Prentice Hall, Upper Saddle River (2003)
8. Freund, Y., Schapire, R.: A decision-theoretic generalization of on-line learning and an application to boosting. Journal of Computer and System Sciences 55(1), 119–139 (1997)
9. Funahashi, T., Fujiwara, T., Koshimizu, H.: Hierarchical tracking of face, facial parts and their contours with PTZ camera. In: 2004 IEEE International Conference on Industrial Technology (ICIT), pp. 198–203. IEEE, Los Alamitos (2004)
10. Hsu, R., Abdel-Mottaleb, M., Jain, A.: Face detection in color images. IEEE Transactions on Pattern Analysis and Machine Intelligence 24(5), 696–706 (2002)
11. Isard, M., Blake, A.: Condensation — conditional density propagation for visual tracking. International Journal of Computer Vision 29(1), 5–28 (1998)
12. Islam, S., Bennamoun, M., Davies, R.: Fast and fully automatic ear detection using cascaded AdaBoost. In: 2008 IEEE Workshop on Applications of Computer Vision, IEEE Workshop on Applications of Computer Vision, pp. 205–210. IEEE, Los Alamitos (2008)
13. Jiang, R.M., Crookes, D.: Multimodal biometric human recognition for perceptual human-computer interaction (draft). IEEE Transactions on Systems, Man and Cybernetics (2010)
14. Kawaguchi, T., Rizon, M., Hidaka, D.: Detection of eyes from human faces by hough transform and separability filter. Electronics and Communications in Japan Part II-Electronics 88(5), 29–39 (2005), doi:10.1002/ecjb.20178
15. Klauser, F.: Interacting forms of expertise in security governance: the example of CCTV surveillance at Geneva International Airport. British Journal of Sociology 60(2), 279–297 (2009), doi:10.1111/j.1468-4446.2009.01231.x
16. Lienhart, R., Maydt, J.: An extended set of haar-like features for rapid object detection. In: Proceedings of IEEE International Conference on Image Processing (ICIP), IEEE Signal Proc. Soc. 2002, vol. I, pp. 900–903. IEEE, Los Alamitos (2002)
17. Lin, J., Ming, J., Crookes, D.: A probabilistic union approach to robust face recognition with partial distortion and occlusion. In: 2008 IEEE International Conference on Acoustics Speech and Signal Processing (ICASSP), vol. 1-12, pp. 993–996. IEEE, Los Alamitos (2008)
18. News, B.: 1,000 cameras 'solve one crime' (August 24, 2009), http://news.bbc.co.uk/1/hi/8219022.stm
19. Nor'aini, A., Raveendran, P.: Improving face recognition using combination of global and local features. In: 2009 6th International Symposium on Mechatronics and its Applications (ISMA), pp. 433–438. IEEE, Los Alamitos (2009)
20. Norris, C.: The Maximum Surveillance Society: the Rise of CCTV. Berg, Oxford (1999)
21. Norris, C., Armstrong, G.: Space invaders: The reality of a CCTV control room in Northern England raises the old question, "Who guards the guards?". Index on Censorship 29(3), 50–52 (2000)

22. Peter, A.: Surveillance at the airport: surveilling mobility/mobilising surveillance. Environment and Planning A 36(8), 1365–1380 (2004), doi:10.1068/a36159
23. Papel, I., Frodel, J.: Facial plastic and reconstructive surgery. Thieme, New York (2002)
24. Pass, A., Zhang, J., Stewart, D.: An investigation into features for multi-view lipreading. In: IEEE International Conference on Image Processing (ICIP). IEEE, Los Alamitos (2010)
25. Register, T.: IT contractors convicted of uk casino hack scam (March 15, 2010), http://www.theregister.co.uk/2010/03/15/uk_casino_hack_scam/
26. Rosen, J.: A cautionary tale for a new age of surveillance (October 7, 2001), http://www.nytimes.com/2001/10/07/magazine/07SURVEILLANCE.html
27. Shapiro, L.G.: Computer vision. Prentice-Hall, Englewood Cliffs (2001)
28. Sharkas, M., Abou Elenien, M.: Eigenfaces vs. Fisherfaces vs. ICA for face recognition; a comparative study. In: ICSP: 2008 Proceedings of 9th International Conference on Signal Processing, vol. 1-5, pp. 914–919. IEEE, Los Alamitos (2008)
29. Sinha, P., Balas, B., Ostrovsky, Y., Russell, R.: Face recognition by humans: Nineteen results all computer vision researchers should know about. Proceedings of the IEEE 94(11), 1948–1962 (2006), doi:10.1109/JPROC.2006.884093
30. Turk, M., Pentland, A.: Face recognition using Eigenfaces. In: 1991 IEEE Computer Society Conference on Computer Vision and Pattern Recognition, pp. 586–591. IEEE, Los Alamitos (1991)
31. Viola, P., Jones, M.: Robust real-time face detection. International Journal of Computer Vision 57(2), 137–154 (2004)
32. Wiskott, L., Fellous, J., Kruger, N., von der Malsburg, C.: Face recognition by elastic bunch graph matching. IEEE Transactions on Pattern Analysis and Machine Intelligence 19(7), 775–779 (1997)
33. Xu, Z., Wu, H.R.: Shape feature based extraction for face recognition. In: ICIEA: 2009 4th IEEE Conference on Industrial Electronics and Applications, vol. 1-6, pp. 3034–3039. IEEE, Los Alamitos (2009)
34. Zhao, W., Chellappa, R.: Face processing (2006), http://www.loc.gov/catdir/enhancements/fy0645/2006296212-d.html
35. Zhao, W., Chellappa, R., Phillips, P., Rosenfeld, A.: Face recognition: A literature survey. ACM Computing Surveys 35(4), 399–459 (2003)
36. Zhou, H., Yuan, Y., Sadka, A.: Application of semantic features in face recognition. Pattern Recognition 41(10), 3251–3256 (2008), doi:10.1016/j.patcog.2008.04.008

Appendix: Further Implementation Details of the Experimental System

The system used to conduct the experiments in this chapter was written in C++ using the OPENCV libraries. The source code is available for free download from our website (http://www.electriceye.org.uk), under the terms of the GNU General Public License. The website also has some short videos illustrating our experimental results.

An overview of the system architecture was described in section 2 and the relationship between front-end camera nodes and the back-end database and classifier was shown in figure 2. Figure 31 shows a more detailed view of a single node and its communication with the back-end. The node detects faces and extracts a feature vector for each face. Object tracking could be carried out by the nodes, so that multiple face samples can be tagged with an identifier, thus attributing the samples to the

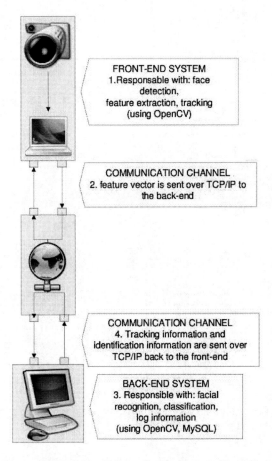

Fig. 31 Distributed System Architecture: Communication between Front-End and Back-End

same person. (This feature was not implemented in our system). The feature vectors are then packaged for transmission to the back-end using TCP/IP.

In section 3.3, we described how the front-end should attempt to match faces to both a frontal and a profile model of the individual. In the design of our system, each match would be performed by an independent thread, as shown in figure 32. The "second thread" in the front-end is in fact two separate threads, one detecting left-facing profiles and one detecting right-facing profiles. Thus, three threads would search for frontal faces, right profiles and left profiles, respectively. A fourth thread handles all the communication with the back-end. (Profile detection was not implemented in our system; see section 3.3.5 for discussion).

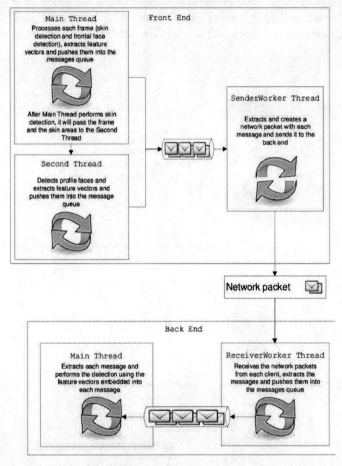

Fig. 32 Front End: Multi-threaded approach to feature extraction and matching against frontal face model and profile face model

Event Understanding of Human-Object Interaction: Object Movement Detection via Stable Changes

Shigeyuki Odashima, Taketoshi Mori, Masamichi Simosaka, Hiroshi Noguchi, and Tomomasa Sato

Abstract. This chapter proposes an object movement detection method in household environments. The proposed method detects "object placement" and "object removal" via images captured by environment-embedded cameras. When object movement detection is performed in household environments, there are several difficulties: the method needs to detect object movements robustly even if sizes of objects are small, the method must discriminate objects and non-objects such as humans. In this work, we propose an object movement detection method by detecting "stable changes", which are changing from the recorded state but which change are settled. To categorize objects and non-objects via the stable changes even though non-objects make long-term changes (e.g. a person is sitting down), we employ motion history of changed regions. In addition, to classify object placement and object removal, we use multiple-layered background model, called the layered background model and edge subtraction technique. The experiment shows the system can detect objects robustly and in sufficient frame-rates.

1 Introduction

Recently, recording daily activities as a lifelog has become feasible due to the wide-spread use of digital cameras and large-scale computer storages. As shown in Fig. 1, object tracking systems that focus on object placement and removal events in lifelogs are possible by using environment-embedded cameras. The object tracking

---

Shigeyuki Odashima · Taketoshi Mori · Masamichi Shimosaka · Hiroshi Noguchi · Tomomasa Sato
Graduate School of Information Science and Technology,
The University of Tokyo, 7-3-1 Hongo, Bunkyo-ku, Tokyo, Japan
e-mail: odashima@ics.t.u-tokyo.ac.jp, tmori@ics.t.u-tokyo.ac.jp
　　　　simosaka@ics.t.u-tokyo.ac.jp, noguchi@ics.t.u-tokyo.ac.jp
　　　　tsato@ics.t.u-tokyo.ac.jp

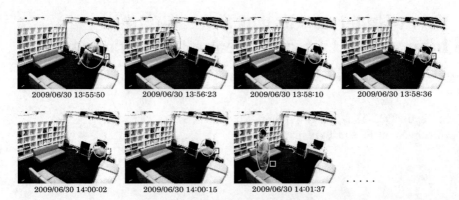

Fig. 1 Object tracking application. Object movements are detected automatically, and users can retrieve their lifelogs of object movements. The key feature of this system is detecting object movement.

system is expected to be helpful in various applications, including alarms that make a sound when items are left behind, support in searching for items, and human action recognition from the interaction of people and objects [1]. In this chapter, we propose a method for image-based event detection of object movement, especially object placement and removal, in a household environment.

There are several difficulties to achieve object movement detection in a household environment. First, the detection method needs robust event detection of object placement and removal. So, the object movement detection method must work robustly for both of object placement and removal even when the sizes of moved objects are small on the input images. Second, the detection method must discriminate between objects and non-objects such as human. The state-of-the-art human detection approaches are mainly appearance-based [2, 3], but especially in the household environment, robust appearance-based human detection is difficult due to occlusions by furniture.

In this chapter, we propose an object movement detection framework via "stable change" of the images. The stable change is the state which is changing from the recorded state, but which change is settled. For example, when a book is placed on a table, the book region is changing from "table", but the region remains as "book". The image changes can be extracted robustly even when the sizes are small by using background subtraction methods, so object detection via the stable changes of the images can detect objects robustly. The proposed method detects the stable changes via temporal information of "motion" of changed regions, to categorize objects and non-objects even though non-objects are occluded and make long-term changes (e.g. a person is sitting down on a chair). To categorize objects and non-objects via temporal information of motion, the proposed method tracks the changed regions and classifies them into objects and non-objects via a state machine driven by motion of the changed regions.

The structure of this chapter is organized as follows. Section 2 provides an overview of our framework. Section 3 describes background modeling and our tracking technique. Section 4 describes the classification method of extracted regions for background updates and object detection. In Section 5, the experimental results show that the proposed object movement detection method works well even when human motions are present. Conclusions are presented in Section 6.

1.1 Related Works

To detect object movement, several types of approaches have been proposed. For example, objects are considered as "highly featured points", so attention point detection techniques are useful to directly detect objects in the images [4, 5]. Approaches based of attention point techniques can be employed even with moving cameras. But, the scene will be less featured when objects are removed, so these methods are difficult especially to detect object removal.

As object detection methods based on background subtraction methods, the method via detection of long-term changes [6] and methods via detection of background model adaptation caused by long-term changes [7, 8] has been proposed. To reject changes caused by human regions, Maki employs a human detection approach based on the size of changed region, color pattern and face detection [6], Harville employs human detection via height of changed regions and tracking [9] and Grabner employs appearance-based human detection [10].

2 Overview of the Proposed Method

Fig. 2 depicts an overview of the proposed method. The proposed method has two major stages: attentive region detection and object detection.

First, the method extracts changed regions by a background subtraction method, and then tracks the extracted regions ("attentive region detection" stage in Fig. 2). In

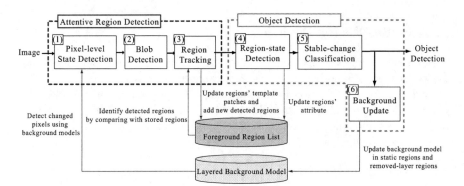

Fig. 2 Overview of the proposed method

this stage, (1) the method extracts changed pixels by a background subtraction technique and categorizes them into "something inserted" state, called the foreground state, and "something removed" state, called removed-layer state. (2)The method then employs the blob detection algorithm to the pixels and extracts changed regions. After extracting changed regions, (3) the method tracks the extracted regions.

Second, the method categorizes the extracted regions into non-objects and objects via their motion, and finally detects object placement and object removal. In this stage, (4) the method discriminates between the non-object state and the object state via the regions' motion detection result for past some frames (we call this motion detection results as "motion history"). (5) The method rejects the stable changes caused without object movements, then detects object movements. Finally, (6) the system updates its background model according to the object detection result.

The proposed method detects object movements from the stable changes extracted by a background subtraction method. Common background subtraction methods have only one background model, so background subtraction methods with only one model have only "changed" state , called foreground state, and "not changed" state, called background state. In this research, we need to classify "object placement" and "object removal" from "changed" state, so we adopt a multiple-layered background model [6, 7] (we call the multiple-layered background model as "layered background model"). Moreover, the proposed method adopts an object placement / removal classification method based on edge subtraction, to handle "object removal which exists in the initial state", which the method with only layered background model cannot handle properly.

The proposed method employs motion history of extracted regions to discriminate between object regions and non-object regions. In the household environment, objects or non-objects are sometimes largely occluded by furniture, so tracking methods must be robust for sudden occlusion. The proposed method adopts keypoint-based tracking method for this problem. In addition, to detect motion robustly even when people move only parts of their body (e.g. when a human is sitting on a sofa and read a book, the whole body of the human hardly move but only their arms moves), we adopt the partial frame subtraction technique.

3 Attentive Region Detection

This section describes the method to extract changed regions from input images and to identify these regions by comparing them to stored regions.

First, we discuss detection method of image changes via a region-level background subtraction technique, then describe how the proposed method categorizes these changes into object placement state and object removal state. Finally, we discuss our tracking method.

3.1 Region-Level Background Subtraction for Object Movement Detection

When the method detects changes caused by object movements, the method needs to be robust for changes caused without object movements (e.g. shadows, background clutters). In this work, we employ a region-level background subtraction method by graph cuts (this method is mainly based on Shimosaka's method [11]).

Background subtraction by using only single pixels fails when background color is similar to foreground object's and is weak for noises. To address these problems, background subtraction methods based on region optimization via graph cuts [11, 12] are proposed. The background subtraction method via graph cut chooses the label x_r of each pixel ($x_r = 0$ if background, $x_r = 1$ if foreground) with optimizing energy function mentioned below:

$$E(X) = \sum_{r \in I} D_r(x_r) + \lambda \sum_{(r,s) \in \varepsilon} S_{rs}(x_r, x_s) \quad (1)$$

where I is the input image and r is the target pixel, ε is the set of adjacent pixel pairs. D_r is the data term, encoding the cost when the label of r is x_r, and S_{rs} is the smoothing term, encoding the cost when the adjacent pixels r,s have different labels. The parameter λ controls the influence of the smoothing term. The energy in eq.(1) is minimized by the min-cut / max-flow algorithm [13].

The data term is defined as follows:

$$D_r(x_r) = \begin{cases} -\ln p_B(i_r) & (x_r = 0) \\ -\ln p_F(i_r) & (x_r = 1) \end{cases} \quad (2)$$

where i_r is RGB vector on pixel r, $p_B(i_r)$ is the background likelihood, $p_F(i_r)$ is the foreground likelihood. We employ robust a color model for shadow changes [14] as the background likelihood. This color model divides RGB difference into the brightness difference and the chromatic difference. The elements of this color model are defined as follows:

$$\begin{aligned} \alpha_r &= \frac{\tilde{i}_r^T i_r}{|\tilde{i}_r|^2}, \\ c_r &= |i_r - \alpha_r \tilde{i}_r|, \end{aligned} \quad (3)$$

where $\tilde{i}_r$ is the RGB vector of pixel r on the background image, α_r is the ratio of the brightness of the input image to the one of the background image on the pixel r, c_r is the chromatic difference.

The background likelihood $p_B(i_r)$ is calculated as follows[11]:

$$p_B(i_r) = \begin{cases} 0 & (\frac{v_r - 1}{\eta_r} < \tau_l \text{ or } \frac{v_r - 1}{\eta_r} > \tau_h) \\ N(c_r | 0, \alpha_r^2 \xi_r^2) & (otherwise) \end{cases} \quad (4)$$

where v_r is $\alpha_r - 1$, $N(c_r | 0, \alpha_r^2 \xi_r^2)$ is Gaussian distribution which mean is 0 and which variance is $\alpha_r^2 \xi_r^2$, η_r is the standard deviation of c_r, ξ_r is the standard deviation of v_r, and τ_l and τ_r are the fixed thresholds.

The foreground likelihood p_F is calculated by using low background likelihood pixels on the input image. To extract changes even when the changes are small, our method employs a histogram foreground likelihood model as follows:

$$p_F(i_r) = \frac{N_B + C}{N_T + K \times C} \tag{5}$$

where N_B is the histogram bin value which includes I_r, N_T is the total histogram bin value, K is the number of bins and C is a constant value (in our implementation, $C = 1$). If N_T is less than the fixed threshold $N_{T_{min}}$, P_F is set to the constant value P_{const} (in our implementation, $N_{T_{min}} = 150$ and $P_{const} = \frac{1}{255^3}$).

The smoothing term S_{rs} uses edge subtraction value of the input image and the background image, to handle both of increase / decrease of edge value caused by object placement / removal:

$$S_{rs} = e^{-\beta |\,||i_r - i_s||^2 - ||\tilde{i}_r - \tilde{i}_s||^2\,|} \tag{6}$$

$$\beta = < \frac{1}{2(||i_r - i_s||^2 + ||\tilde{i}_r - \tilde{i}_s||^2)} > \tag{7}$$

where β is a parameter that balances the color contrast, $||\cdot||^2$ is the L_2 norm and $<\cdot>$ is the expectation operator.

3.2 Pixel-Level State Detection by Layered Background Model

After the proposed method extracts changed pixels from the input images, then the method categorizes the changed pixels into the foreground state and the removed-layer state. To categorize pixels into these three state, background state (unchanged), foreground state and removed-layer state, we employ the layered background model. Fig. 3 depicts an overview of our layered background model. The layered background model contains two background models: the base background and the layered background. The base background records static backgrounds (e.g. furniture), and the layered background overlays placed objects on the base background. The method generates the base background when object movement detection starts (in this chapter, we call the state before the object movement detection starts as "initial state"). The method inserts the detected object into the layered background when the method detects object placement, and delete the detected object from the layered background when the method detects object removal.

Pixel categorization by the layered background model is performed as follows. First, the method compares the input image to the layered background, and extracts changed pixels by the region-level background method. Next, the method compares the changed pixels in the input image to the base background. if the pixel is changing from the layered background but not changing from the base background, the pixel is changing after object placement but is not changing before object placement, so it represents "something removed". So, the method classifies the pixel as the removed-layer state. On the other hand, the pixel is changing from both of the

Object Movement Detection via Stable Changes 201

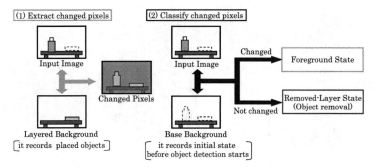

Fig. 3 Layered background model. Each pixel on the input image is compared to the layered background and the base background. First, the input image is compared to the layered background, and changed pixels on the input images are extracted. Second, the changed pixels are compared to the base background, and then these changed pixels are classified into the foreground state or the removed-layer state

layered background and the base background, the pixel is classified as the foreground state.

Classification of object placement and object removal via the layered background model can detect easily which object in the detected objects is removed. But, if objects which exists in the initial state are removed, the regions of the removed objects change from both of the layered background and the base background, so the pixels of the removed object are classified as the foreground state. Therefore, only with the layered background model, removal of objects in the initial state can't be handled. To solve this problem, we employ edge subtraction technique. Details of the classification by edge subtraction technique is discussed in Section 4.2.

After pixel-level state detection, the blob detection algorithm is employed for foreground state pixels and removed-layer state pixels. Blobs of foreground state pixels whose size is more than R_{th} pixels are extracted as the foreground region, and blobs of removed-layer state pixels whose size is more than R_{th} pixels are extracted as the removed-layer region.

Foreground regions extracted in this operation are then tracked. The size of the removed-layer region is checked in the background update process. If the ratio of the size of the removed-layer region to the size of its original layered background is greater than a fixed threshold, object removal is detected. In our implementation, the threshold ratio is set to 0.80.

3.3 Region Tracking

After the method extracts changed regions, then the method tracks the extracted regions. To track robustly under occlusion, we apply keypoint-based tracking approach via FAST-10 corner detector [15]. Fig. 4 shows an overview of keypoint based tracking and the detection result.

Fig. 4 Keypoint based tracking. (a): Overview of keypoint based tracking. Extracted regions are matched with the list of stored foreground regions by using their keypoint patch templates. (b): Tracking result of human using keypoint based tracking. The circles represent extracted patch template area

First, the method extracts the FAST keypoints in the inner foreground regions. To detect keypoint patches in sufficient scales, the method generates a five-level image pyramid, and then extracts keypoints in each level. Level zero of the image pyramid stores the full resolution image, and images in the image pyramid are sub-sampled in each level. The method extracts keypoints from grayscale-converted images on each image level and generates patch templates of 5×5 pixel for these keypoints from RGB images on the extracted level. To be robust under illumination changes, each RGB pixel is converted to a color invariants [16, 17].

The patch templates are compared to the patch templates of the foreground regions extracted in previous frames, which are stored in the system's foreground region list. To reduce the computational cost, only patches in the same pyramid level are compared. This comparison is done by evaluating the sum-of-squared-differences (SSD) score at all keypoints within the search regions and selecting the keypoint with smallest difference score. If the best matching score is under the fixed threshold (0.05 in our implementation) and if the ratio of the best matching score to the second-best matching score is less than the fixed threshold (0.8 in our implementation), the patch is considered to be matched.

Up to n matched points (10 in our implementation) are searched in the extracted foreground region, and the extracted region is identified as the region with the largest number of matched points in the foreground region list. To avoid tracking failure caused by wrong keypoint matching, if the ratio of each region size is over a fixed threshold (10 in our implementation), the regions are regarded as not matched. Additionally, to track very low textured changed regions (e.g. when objects in the initial state are removed, the changed regions are usually the part of low textured furniture), if the distance of the center of each region is small (2 pixel in our implementation) and the ratio of each region size is under the threshold mentioned above, the regions are matched.

If an extracted region does not match any regions, then it is added to the foreground region list. In some cases, several extracted regions match to one stored region in the foreground region list (e.g. when a person places a book on something, the human region and the book region will match the region of the human with the book in a previous frame). In these cases, the human region is expected to be a larger object region. So, the larger region is identified as the matched region, and the smaller region is newly registered to the foreground region list. If the region stored in the foreground region list is not accessed for a long time, the region is deleted.

4 Object Detection and Background Update

This section describes the technique for categorizing the extracted regions into non-objects and objects via their motion history, and detecting object placement and object removal from foreground regions' edge subtraction result.

4.1 Region State Detection

After the method tracks extracted regions, the method detects extracted regions' motion via the partial frame subtraction technique. Fig. 5 (a) shows an overview of partial frame subtraction. In the partial frame subtraction method, the extracted region is divided into $M \times N$ subregions (4×4 in our implementation), and motion is detected in each subregions using the frame subtraction technique. If the ratio of pixels detected moving in the subregion are above a fixed threshold, the whole extracted region is classified as a moving region. Partial human motion (e.g. arm movement) is expected to form a cluster of moving state pixels, so partial frame subtraction enables to robust detection of partial motion (an example is shown in Fig. 5 (b)).

Then the method updates the attribute of the extracted foreground regions by their motion detection history. Each foreground region has stability value $S(t)$, where k is an incremental parameter. The stability values are updated by the region's motion detection result as follows.

1. If the region state is moving, $S(t) = S(t-1) + k$
2. If the region state is stable, $S(t) = S(t-1) - k$

An unstable result of motion detection may cause misclassification of the region's attribute. To avoid detecting attributes under unstable motion detection, we set incremental parameter $k = 2$ if the region's motion detection result is equal to the previous result, and $k = 1$ if the region's motion detection result is not equal to the previous result.

After updating the stability value, then the method calculates the region's attribute via a state machine driven by stability value $S(t)$. Fig. 6 shows the state diagram of the region's attribute. Three attributes are defined as follows.

1. **Transition region**
   The initial attribute of a foreground region is the transition region. If the region with transition region attribute does not transfer to the update-prohibited region and the static region, the region transfers to the transition region.

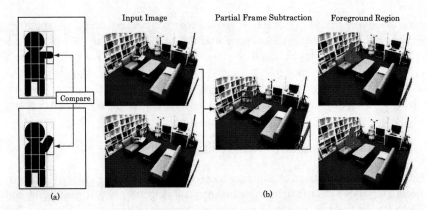

Fig. 5 Partial frame subtraction. (a) An overview of partial frame subtraction. (b) Motion detection result via partial frame subtraction. The left images are the input images, the center image is the partial frame subtraction result, and the right images represent the foreground regions. In the center images of (b), rectangles are overlaid on regions where the method detects motion: the outer rectangles represent that the method detects the foreground's region is moving via the partial frame subtraction, the inner rectangles are the partial regions which the method detects via the proposed method. When you focus on the human region, the contours of the regions are almost the same, but the method detects movement of human arm by the partial frame subtraction.

2. **Update-prohibited region** (object state)
   If the stability value of the region in the transition region is over some threshold S_{th} ($S(t) > S_{th}$), it transfers to the update-prohibited region. This attribute represents the region is non-object because the region was moving for a long time. To avoid classifying long-term change caused by non-objects as placed objects, the update-prohibited regions transfer only to the update-prohibited regions.
3. **Static region** (non-object state)
   If the stability value of the region with the transition region is $S(t) < -S_{th}$, it transfers to the static region. In the background update stage, static regions are deleted from the foreground region list and are inserted to the layered background.

In our implementation, we set the threshold parameter $S_{th} = 20$.

4.2 Stable-Change Classification

After the method extracts static regions, the method rejects the static regions without object movements (e.g. shadows, small object shift), then the method classify the object regions into object placement and object removal. To reject the stable changes without object movements, we set some threshold parameters - Bhattacharrya distance between HSV histogram (we employed Pérez's HSV histogram [18] in our method) of the input image and the background image in the extracted regions

Fig. 6 Region-state update by the state machine. The attribute transfer from the transition region to the update-prohibited region or the stable region when the stability value is greater than S_{th} and less than $-S_{th}$, respectively. The regions classified as the update-prohibited region never transfer to another attribute. The regions classified as the stable region are deleted from the foreground region list and are inserted to the layered background.

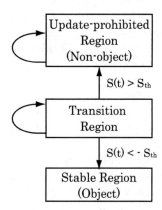

(C_{th}), the ratio of the major axis to the minor axis of the region by approximating the region to ellipse (E_{rth}) and the length of the minor axis of the region (L_{mth}), and the average width of the region (W_{th}). Bhattacharrya distance $d_h(q(E_i), q(E_b))$ between a histogram $q(E_i)$ and a histogram $q(E_b)$ is calculated as follows.

$$d_h(q(E_i), q(E_b)) = \sqrt{1 - \sum_{k=1}^{N} \sqrt{q(E_i;k)q(E_b;k)}} \qquad (8)$$

As mentioned in Section 3.2, the regions where objects are placed and the regions where initial objects are removed are classified into foreground state. To distinguish this two regions, we employ a classification method based on edge subtraction [19].

Generally, the region where objects don't exist is less textured than where objects exist. So, textures of the region will increase when objects are placed, and will decrease when objects are removed. The method extracts the amount of edge energies in the boundary of the stable region of the input image and the layered background, and if the energy of the input image is greater than the one of the layered background, the method classifies the region as object placement. Otherwise, the method classifies the region as object removal. The method determines object placement and object removal via the edge difference measure E_c as defined below:

$$E_c = \sum_{r \in R_c} \frac{1}{|s|} \sum_{s} (||i_r - i_s||^2 - ||\tilde{i}_r - \tilde{i}_s||^2) \qquad (9)$$

where R_c is the contour of the extracted region, i_r and $\tilde{i}_r$ are the RGB vector of pixel r on the input image and those on the background image, respectively. s is the 4-neighbor pixels of the pixel r. If $E_c > 0$, the region is determined as object placement, and if $E_c < 0$, the region is determined as object removal.

4.3 Object Detection and Background Update

By using the detection result of the extracted foreground regions and removed-layer regions, the method detects object placement by detecting static regions which the input image's edge boundary energy is larger than the layered background's. In addition, the method detects object removal by detecting removed-layer regions and static regions which the input image's edge boundary energy is smaller the layered background's.

Then the method update the background model according to detection result. Fig. 7 depicts an overview of background update. When the method detects placed objects, the method inserts the detected objects to the layered background. When the method detects removed objects by the layered background model, the method deletes the detected objects from the layered background. When the method detects removed objects by edge subtraction, the method updates the base background of the detected object regions.

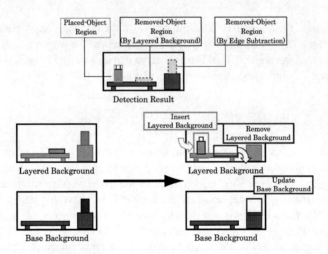

Fig. 7 Overview of background update. The base background and the layered background are updated according to object detection results

5 Results

5.1 Experiments

We evaluate the object movement detection performance of proposed method with 8 video sequences which are captured in 4 viewpoints (total 32 different video sequences, total 10324 frames). These video sequences consist of images of 320×240 resolution recorded at 7.5 fps. The video sequences contain 94 object placement and 58 object removal (total 152 events).

The object detection method was implemented on a PC with an Intel Core 2 Duo 2.5 GHz processor. The method ran with single-thread processing.

In this experiment, we implement false positive and recall as performance evaluation measures, as defined below:

$$\text{false positive} = 1 - \frac{\text{correctly detected object movements}}{\text{total detected object movements}} \quad (10)$$

$$\text{recall} = \frac{\text{correctly detected object movements}}{\text{total object movements in the images}} \quad (11)$$

In this experiment, we calculated performance of the proposed method under variant threshold parameters $R_{th}, C_{th}, E_{rth}, L_{mth}, W_{th}$. We compare the proposed method with the proposed method without the layered background model (object movement detection by only edge subtraction), and our previous method by the pixel-level layered background model [20]. The original object movement detection method via the pixel-level layered background model cannot handle object movements which exists in initial state, we added edge subtraction based object movement classification (mentioned in 4.2) to the method.

Fig. 8 shows the resulting detection performance in various parameters. As can be seen from the graph, where false positive is from 0.05 to 0.25, the detection performance by the region-level background subtraction method is superior to the detection performance by the pixel-level background subtraction method. This is because of that the region-level background subtraction method is robust to background clutter (e.g. small object shift, small shadow regions).

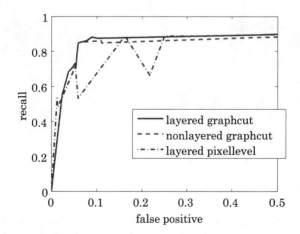

Fig. 8 ROC curves of the proposed method. The solod line: by the proposed method. The broken line: by the proposed method without the layered background model. The dashed-dotted line: by our pixel-level object movement detection method [20].

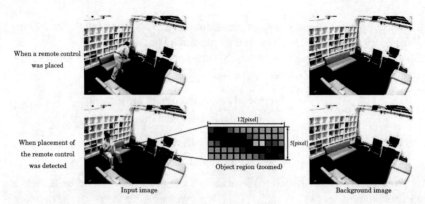

Fig. 9 Detection result of placing object. Top: the moment when a person places a remote control. Bottom: the moment when the proposed method detects the remote control. Left: the input image, middle: the detected object region (zoomed), right: the background image. In the left-bottom image, placement of the remote control was detected appropriately, and at the same time, in the right-bottom image, the remote control region was added to the background model.

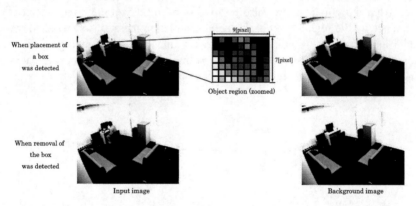

Fig. 10 Detection result of removing object. Top: the moment when the proposed method detects placement of a box. Bottom: the moment when the proposed method detects removal of the box. Left: the input image, middle: the detected object region (zoomed), right: background image. In the right-bottom image, removal of the box was detected appropriately, and at the same time, the layered background of the box was removed.

Fig. 9 and Fig. 10 show object detection results by the proposed method (these results are taken with threshold parameters when false positive $= 0.08$ and recall $= 0.86$). Fig. 9 shows the detection result of the sequence when a person places a remote control on the sofa and then sits down on the sofa. In this sequence, the remote control is detected as a placed object, but the person is not detected as an object. The passed time from the remote control placement and the person sitting is almost the same, but the system detects only the remote control as a placed object, so the proposed method works correctly. At the same time, the size of the remote control is small (only 12 × 5[pixel]), the system successfully detects object movements. Fig. 10 shows the detection result of the sequence when a person places a box on the table, sits down on the chair and then removes the box. In this case, even the person is sitting down and his body is occluded by a chair, the method does not detect the person as an object but detects object removal. So, the motion-based classification method of object and non-object works robustly.

The average process time of the proposed method was 130[ms/frame], so the proposed method works in sufficient frame-rates.

5.2 Limitations

The proposed method has some limitations. First, the background subtraction method can't handle strong illumination changes (e.g. switching off the lights or opening a curtain). Second, the proposed method can't handle movement of furniture (e.g. opening and closing of a door, large shift of chair) because movement of furniture is neither of "object placement" and "object removal", that the proposed method can handle. Third, the proposed method hardly detect object movements when the color of the object is very similar to near object's (e.g. when a white telephone is placed near white curtains). In this case, robust object movement detection is difficult by using images captured only in a single viewpoint. So, to detect objects more robustly, the method needs to integrate multiple viewpoint information.

6 Conclusion

This chapter proposed an object movement detection method in a household environment via the stable changes of images. To detect object placements and object removals robustly, the method employs the layered background model and the edge subtraction based classification method. In addition, to classify objects and non-objects robustly though the changed regions are occluded, the method uses motion history of the regions.

Our experiment shows the method detects objects robustly even though the object size on the image is small, and the method distinguishes between object placement and removal appropriately. The method runs in sufficient frame rates, so the proposed method works well.

Future tasks are handling large object shift such as movement of furniture, integrating multiple viewpoints to detect objects more robustly.

References

1. Gupta, A., Kembhavi, A., Davis, L.: Observing human-object interactions: using spatial and functional compatibility for recognition. IEEE Transactions on Pattern Analysis and Machine Intelligence 31(10), 1775–1789 (2009)
2. Navneet, D., Triggs, B.: Histograms of oriented gradients for human detection. In: CVPR (2005)

3. Sabzmeydani, P., Mori, G.: Detecting pedestrians by learning shapelet features. In: CVPR (2007)
4. Itti, L., Baldi, P.: A principled approach to detecting surprising events in video. In: CVPR (2005)
5. Gould, S., Arfvidsson, J., Kaehler, A., Sapp, B., Messner, M., Bradski, G., Baumstarck, P., Chung, S., Ng, A.: Peripheral-foveal vision for real-time object recognition and tracking in video. In: IJCAI (2007)
6. Maki, K., Katayama, N., Shimada, N., Shirai, Y.: Image-based automatic detection of indoor scene events and interactive inquiry. In: ICPR (2008)
7. Kim, K., Chalidabhongse, T., Harwood, D., Davis, L.: Real-time foreground–background segmentation using codebook model. Real-Time Imaging 11(3), 172–185 (2005)
8. Tian, Y., Lu, M., Hampapur, A.: Robust and efficient foreground analysis for real-time video surveillance. In: CVPR (2005)
9. Harville, M.: A framework for high-level feedback to adaptive, per-pixel, mixture-of-gaussian background models. In: Heyden, A., Sparr, G., Nielsen, M., Johansen, P. (eds.) ECCV 2002. LNCS, vol. 2352, pp. 543–560. Springer, Heidelberg (2002)
10. Grabner, H., Roth, P., Grabner, M., Bischof, H.: Autonomous learning of a robust background model for change detection. In: Performance Evaluation of Tracking and Surveillance (PETS) Workshop at CVPR (2006)
11. Shimosaka, M., Murasaki, K., Mori, T., Sato, T.: Human shape reconstruction via graph cuts for voxel-based markerless motion capture in intelligent environment. In: IUCS (2009)
12. Sun, J., Zhang, W., Tang, X., Shum, H.-Y.: Background cut. In: Leonardis, A., Bischof, H., Pinz, A. (eds.) ECCV 2006. LNCS, vol. 3952, pp. 628–641. Springer, Heidelberg (2006)
13. Boykov, Y., Kolmogorov, V.: An experimental comparison of min-cut/max-flow algorithms for energy minimization in vision. IEEE Transactions on Pattern Analysis and Machine Intelligence 26(9), 1124–1137 (2004)
14. Horprasert, T., Harwood, D., Davis, L.: A robust background subtraction and shadow detection. In: ACCV (2000)
15. Rosten, E., Drummond, T.: Machine learning for high-speed corner detection. In: Leonardis, A., Bischof, H., Pinz, A. (eds.) ECCV 2006. LNCS, vol. 3951, pp. 430–443. Springer, Heidelberg (2006)
16. Gevers, T., Smeulders, A.: PicToSeek: combining color and shape invariant features for image retrieval. IEEE Transactions on Image Processing 9(1), 102–119 (2000)
17. Nguyen, H., Smeulders, A.: Fast occluded object tracking by a robust appearance filter. IEEE Transactions on Pattern Analysis and Machine Intelligence 26(8), 1099–1104 (2004)
18. Pérez, P., Hue, C., Vermaak, J., Gangnet, M.: Color-based probabilistic tracking. In: Heyden, A., Sparr, G., Nielsen, M., Johansen, P. (eds.) ECCV 2002. LNCS, vol. 2350, pp. 661–675. Springer, Heidelberg (2002)
19. Connell, J., Senior, A., Hampapur, A., Tian, Y.L., Brown, L., Pankanti, S.: Detection and tracking in the IBM peoplevision system. In: ICME (2004)
20. Odashima, S., Mori, T., Shimosaka, M., Noguchi, H., Sato, T.: Object movement event detection for household environments via layered-background model and keypoint-based tracking. In: International Workshop on Video Event Categorization, Tagging and Retrieval (2009)

Survey of Dirac: A Wavelet Based Video Codec for Multiparty Video Conferencing and Broadcasting

Ahtsham Ali, Nadeem A. Khan, Shahid Masud, and Syed Farooq Ali

Abstract. The basic aim of this book chapter is to provide a survey on BBC Dirac Video Codec. That survey, would not only provide the in depth description of different version of Dirac Video Codec but would also explain the algorithmic explanation of Dirac at implementation level. This chapter would not only provide help to new researchers who are working to understand BBC Dirac video codec but also provide them future directions and ideas to enhance features of BBC Dirac video codec.

Compression is significantly important because of being bandwidth limited or expensive for widespread use of multimedia contents over the internet. Compression takes the advantage in limitation of human perception due to which it is not able to process all the information of perfectly reproduced pictures. We can compress the pictures without the loss of perceived picture quality. Compression is used to exploit the limited storage and transmission capacity as efficiently as possible.

The need of efficient codec's has gained significant attraction amongst researchers. The applications of codec's range from compressing high resolution files, broadcasting, live video streaming, pod casting, and desktop production. Depending on the type of application, the requirements of the codec's change.

1 Background of Codecs

Video Codecs broadly fall under two categories. One is based on the DCT Transform and the other one on the Wavelet Transform. As the DCT transform

---

Ahtsham Ali . Nadeem A. Khan . Shahid Masud
Dept. of Computer Science, LUMS, Lahore, Pakistan
e-mail: ahtsham.ali@lums.edu.pk, nkhan@lums.edu.pk,
　　　smasud@lums.edu.pk

Syed Farooq Ali
Dept. of Computer Science, Ohio State University, USA
e-mail: ali.215@buckeyemail.osu.edu

based codec's are the most popular, in Figure 1 the DCT based encoder block diagram has been shown. The three main building blocks are DCT, quantization and VLC coding. For intra block, the block is DCT transformed, quantized and then VLC coded. There is no need to keep the reference of intra coded block. For inter block, motion vectors are achieved by performing motion estimation. The motion vectors are applied on the reference block, stored in the frame store. The difference block of the current and motion compensated block is DCT transformed, quantized and VLC coded. The motion vectors are also VLC coded and are sent along with difference block coded data. The difference block data after quantization is inverse quantized and inverse DCT transformed. The motion compensated block is also added to this block and then stored in the frame store for future reference.

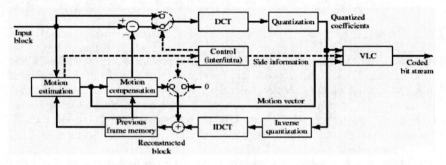

Fig. 1 DCT encoder block diagram [11]

In Figure 2, DCT decoder block diagram has been shown. For intra block, the coded bit stream is inverse VLC coded, inverse quantized and then inverse DCT transformed for reconstruction of block. For inter block, the coded bit stream is inverse VLC coded. The motion compensation is performed on the reference block by applying motion vectors, stored in the previous frame memory. The difference block after inverse VLC coded is inverse quantized and IDCT transform and then added to the motion compensated block to reconstruct the block. The reconstructed block is stored in the previous frame memory for future reference.

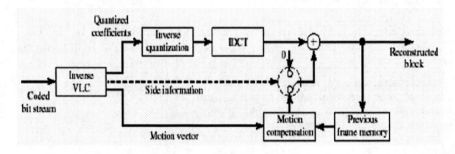

Fig. 2 DCT decoder block diagram [11]

Ghanbari et. al. in [2] gives the background and history of various coding standards. The first video-coding standard developed by CCITT (now ITU-T) in 1984 was H.120. Shortly, afterwards, a new standard H.261 was developed and then H.120 was abandoned. H.261 becomes the originator of all modern video compression standards. Core of all the standards from H.261 onwards is almost same and is as shown in Figure 1 and Figure 2. Some salient features of this codec are the following:

- 16×16 MB motion estimation/compensation
- 8×8 DCT
- Zig-zag scanning of DCT coefficients
- Scalar quantization of those coefficients, and subsequent variable length coding (VLC)
- Loop filter
- Integer-pixel motion compensation accuracy
- 2-D VLC for coding of coefficients

Later the H.261 was overtaken by H.263.

The first coding standard developed by ISO for motion pictures was MPEG-1 (1991). MPEG-1 provides the following features:

- Bi-directional prediction for B-pictures
- Half-pixel motion estimation
- Better quality at higher bit rates

ISO and ITU-T jointly developed a coding standard MPEG-2 in the period 1994/95. MPEG-2 supports two new features that are as follows:

- Interlaced scan pictures
- Scalability

In 1995, H.263 was first developed to replace H.261 as the dominant video conferencing codec. H.263 supports following features:

- Half-pixel motion compensation
- 3-D VLC of DCT coefficients
- Median motion vector prediction
- Optional enhanced modes such as increased motion vector range, advance prediction mode with Overlapped Block Motion Compensation (OBMC)
- Arithmetic entropy coding
- Coding of PB frames to reduce coding overhead at low bit-rates.

MPEG-4 developed in early 1999 follows the H.263 design. Besides including all prior features, it also has the following features:

- Zero-tree wavelet coding of still pictures
- Dynamic 2D mesh coding of synthetic objects and facial animation modeling
- Quarter-pixel motion compensation and global motion compensation

Ghanbari et. al. in [2] further says that H.264 is a state-of-the-art video codec, standardized in 2003 and is suitable for a wide range of applications. This codec has the following features:

- High profile for HDTV and Blue-ray disc storage support a wide set of applications.
- Baseline profile is intended for applications with limited computing resources, such as video-conferencing and mobile applications.
- The main profile was intended for broadcast TV and storage
- The Extended profile is intended for streaming applications, with robust coding, trick modes, and server switching.
- Fixed-point implementation and the first network friendly coding standard.
- Higher computational complexity but better coding efficiency than previous standards.
- An increased range of quantization parameters, and employs Lagrangian optimized rate control.
- The components of the target bit-rate in a rate-distortion function are divided between the individual coding parts in such a manner that maximum reduction in distortion is achieved at the expense of minimal increase in the bit rate.

Wavelet based coding has been used in JPEG 2000 and MPEG-4 for coding images. However, there is no Wavelet based video coding standard yet. Wavelet-based video encoders have attracted significant attention amongst researchers recently, due to the arrival of efficient quantization algorithms and associated coders [2]. Ghanbari et al further says in [2] that the principal attraction of wavelet transform is its inherent property of transformation that lends itself to scalable coding at entropy coder. Scalability is a natural part of the wavelet based codec, in that the coding gain increases at higher scalability levels. However, in DCT-based codecs, larger levels of scalability reduce its compression efficiency. The wavelet may also outperform the DCT in compactness.

Ghanbari et al in [2] also adds that the wavelet-based video coding has come to prominence and will be part of future codecs due to increased demand for scalability in video coding. In Wavelet Transform, more specifically, Haar transform is used for scalable coding between pictures (not for residual MBs) in H. 264. Haar transform also exhibits the properties of the orthogonal transforms.

Ghanbari et al in [2] tells about another advantage of wavelet transform. It comes with the fact that it is applied to a complete image rather than a block. The disadvantage of block based processing is the discontinuities that occur at block edges. In case of H.264, a de-blocking filter is applied at the decoder to resolve discontinuities.

The wavelet transform has been widely applied to still images as well. It is an observation that the Discrete Wavelet Transform (DWT) outperforms a DCT by an average 1dB [9]. Wavelet transform was selected in JPEG2000 and MPEG-4 due to its better energy clustering properties and it is also frame based [2]. Different error resilient algorithms have also been proposed that are based on partitioning of Wavelet Transform Coefficients [15].

2 Advantages of Video Event Analysis in Wavelet

Video Data coded using wavelet transform offer several new avenues to video event analysis.

Video event analysis is generally carried out on raw (un-coded) video data. As video data is large it is usually available in coded form and requires decoding first. As wavelet transformed image data also contain a low resolution image it offers avenues for direct video analysis on this image, hence offering speed advantages. Similar analysis can be contemplated for the intra coded wavelet video.

Wavelet transform enables the extraction of multiple resolution images from a single coded data stream. This makes it easy to extract a lower resolution or lower quality video from the compressed code-stream. This is specifically true for coding paradigm where the wavelet transform is applied to the complete images rather than on image blocks. This offers avenue for performing multi-resolution image analysis leading to identification of multi-resolution events in video. We will therefore, discuss how scalable video can be obtained in case of Dirac after discussing the architecture of Dirac in detail.

3 BBC Dirac Video Codec

Dirac is an experimental Open Source video codec initially developed by BBC Research. It is a royalty-free video codec, named in the honor of the British Scientist Paul Dirac. The aim of the Dirac was to build up a high performance video compression codec with a much simpler and modular design both conceptually and in implementation. Dirac was required to be able to encode and decode the videos in real-time, however, it is too slow for real-time conferencing in its current state. Dirac documentation including high level algorithmic description, doxygen-based comments on the reference code, specification of Dirac bit stream and decoding are provided with Dirac video codec. Dirac is also a wavelet based codec. This book chapter discusses the architecture, features and algorithmic description of Dirac in detail. It also highlights different directions of ongoing and future research. In [10], different approaches for optimizing Dirac are also described.

4 Overview of Dirac Video Codec

4.1 Dirac Architecture

Dirac is an open source video codec developed by BBC for video sizes ranging from QCIF to HDTV progressive and interlaced. The major difference between Dirac and other video codec is the transform coding algorithm. Dirac uses Wavelet Transform instead of DCT Transform.

4.2 Dirac Encoder

In Figure 1, Dirac encoder architecture has been shown. It is very simple and modular. There are four major modules of Dirac including Wavelet Transform, Motion Estimation and Compensation, Entropy Coding and Quantization. Motion Estimation and Compensation module uses Hierarchical Motion Estimation and Compensation. Hierarchical Motion Estimation and Compensation is achieved by down converting each frame of video by a factor of 2, hence finding motions vector at different levels. Dirac performs Integral Error Calculations, Langrangian Calculations and based on these calculations, it selects the best quantizer from 49 quantizers. Dirac uses Arithmetic coding algorithm, but this algorithm is also context based and adaptive in nature. It first converts the number into binary form and based on the context of binary symbol it also updates its probability table.

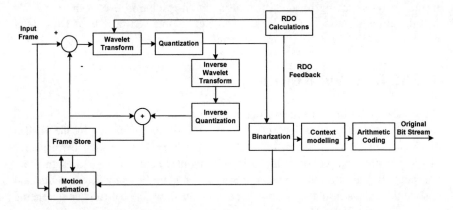

Fig. 3 Dirac encoder architecture

4.3 Dirac Decoder

In Figure 2, Dirac decoder architecture has been shown which performs the inverse of encoder. The decoding process is carried out in three stages. In the first stage, encoded bit-stream is decoded using entropy decoding, the next inverse scaling and quantization is performed and after this inverse transform is performed.

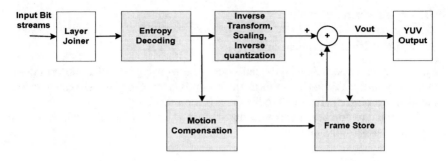

Fig. 4 Dirac decoder architecture

5 Major Modules of Dirac

At this time, Dirac supports planar YUV format streams. YUV is a popular way of representing color images. It separates the brightness information (Y) from the color information (U and V). The Y component is a weighted average of the three additive primaries - red, green and blue. The U component is the difference between the blue primary and the Y component. The V component is the color difference between the red primary and Y component. All three color primaries can be reconstructed from the Y, U and V components. Because the human eye is more sensitive to brightness than color, the chroma components generally are recorded with fewer samples than the brightness data.

The supported YUV formats are:

Table 1 Different Supported Formats

Chroma Format	YUV format Description
format444	Planar 4:4:4 format. Full chroma resolution in both vertical and horizontal directions.
format422	Planar 4:2:2 format. Full vertical chroma resolution and 1/2 horizontal chroma resolution.
format420	Planar 4:2:0 format. 1/2 chroma resolution in both vertical and horizontal directions.

Dirac does not support any other input formats currently. The utility conversion directory has a number of utilities to and from Dirac formats to other formats like RGB and BMP.

5.1 Quantization

5.1.1 Dead-Zone Quantization

Each subband's coefficients are quantised by Dirac using so-called uniform dead-zone quantisers. A simple uniform quantiser is a division of the real line into equal-width bins, of size equal to the quantization factor Q_f:

The bins are numbered and a reconstruction value is selected for each bin. So the bins consist of the intervals.

$$[\left(N-\frac{1}{2}\right)Q_f, \left(N+\frac{1}{2}\right)Q_f]$$

The labels for the bin are integer N, that is subsequently encoded. The reconstruction value used in the decoder (and for local decoding in the encoder) can be any value in each of the bins. The obvious, but not necessarily the best, reconstruction value is the midpoint NQ_f.

A uniform dead-zone quantizer is slightly different. Bins in that quantizer containing zeros is twice as wide. So the bins consist of [-Qf ,Qf] , with a reconstruction value of 0, together with other bins of the form.

$$[\left(N-\tfrac{1}{2}\right)Q_f, \left(N+\tfrac{1}{2}\right)Q_f] \text{ for N>0 \&}$$
$$[(N-1)Q_f, (N)Q_f] \quad \text{for N<0 with reconstruction points somewhere in the intervals.}$$

5.1.1.1 Advantages of Dead Zone Quantiser

The advantage of the dead-zone quantiser is two-fold. It applies more severe quantisation of the smallest values, which acts as a simple but effective de-noising operation. Secondly, it admits a very simple and efficient implementation: simply divide by the quantisation factor and round towards zero. In Dirac, this process is approximated by a multiplication and a bit shift.

A value of X=0.5, giving the mid-point of the interval might be the obvious reconstruction point, as it gives the mid-point of the bin. This is indeed what we use for intra pictures. For inter pictures (motion compensated prediction residues), the values of transformed coefficients in a wavelet subband have a distribution with mean very near zero and which decays pretty rapidly and uniformly for larger values. Values are therefore more likely to occur in the first half of a bin than in the second half and the smaller value of X=0.375 reflects this bias, and gives better performance in practice [8].

5.1.2 Lagrangian Parameter Control of Subband Quantization

Selection of quantisers is a matter for the encoder only. The current Dirac encoder uses an RDO technique to pick a quantiser by minimising a Lagrangian combination of rate and distortion. Essentially, lots of quantisers are tried and the best picked. Rate is estimated via an adaptively-corrected measure of zeroth-order entropy measure Ent(q) of the quantised symbols resulting from applying the quantisation factor q, calculated as a value of bits/pixel. Distortion is measured in terms of the perceptually-weighted error fourth-power error E(q,4), resulting from the difference between the original and the quantised coefficients.

5.2 Wavelet Transform

After the motion compensation is performed, the residuals of motion-compensation are treated almost identically to Intra picture data. In both cases, we have one luminance and two chrominance components in the form of two-dimensional arrays of data values. There are three stages of picture (frame or field) component data as follows:

The data arrays are wavelet-transformed using separable wavelet filters and divided into subbands.

Then they are quantized using RDO quantizers in the reference encoder.

Finally the quantized data is entropy coded.

The architecture of coefficient coding is shown here:

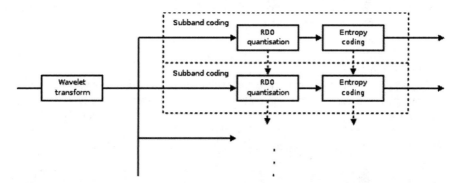

Fig. 5 Coefficient coding architecture [8]

As can be seen, each wavelet subband is coded in turn. Both the quantisation and the entropy coding of each band can depend on the coding of previously coded bands. This does limit parallelisation, but the dependences are limited to parent-child relationships, so some parallelisation/multi-threading is still possible. The only difference between Intra picture coefficient coding and Inter picture residual coefficient coding lies in the use of prediction within the DC wavelet subband of Intra picture components.

At the decoder side, the three stages of the coding process are reversed. The entropy coding is decoded to produce the quantised coefficients, which are then reconstructed to produce the real values. Then, after undoing any prediction, the inverse transform produces the decoded picture component. The reference Dirac encoder has to maintain a local decoder within it, so that compressed pictures must be used as reference frames for subsequent motion compensation else the encoder and the decoder will not remain in sync.

The discrete wavelet transform is now extremely well-known and is described in numerous references. In Dirac it plays the same role of the DCT in MPEG-2 in decorrelating data in a roughly frequency-sensitive way, whilst having the advantage of preserving fine details better. In one dimension it consists of the iterated application of a complementary pair of half-band filters followed by sub sampling by a factor 2:

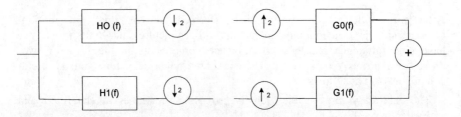

Fig. 6 Perfect reconstruction analysis and synthesis filter pairs [8]

These filters are termed the analysis filters. Corresponding synthesis filters can undo the aliasing introduced by the critical sampling and perfectly reconstruct the input. Clearly not just any pair of half-band filters can do this, and there is an extensive mathematical theory of wavelet filter banks. The filters split the signal into a low frequency (LF) and a high frequency (HF) part; the wavelet transform then iteratively decomposes the LF component to produce an octave-band decomposition of the signal.

Applied to two-dimensional images, wavelet filters are normally applied in both vertical and horizontal directions to each image component to produce four so-called subbands termed Low-Low (LL), Low-High (LH), High-Low (HL) and High-High (HH). In the case of two dimensions, only the LL band is iteratively decomposed to obtain the decomposition of the two-dimensional spectrum shown below:

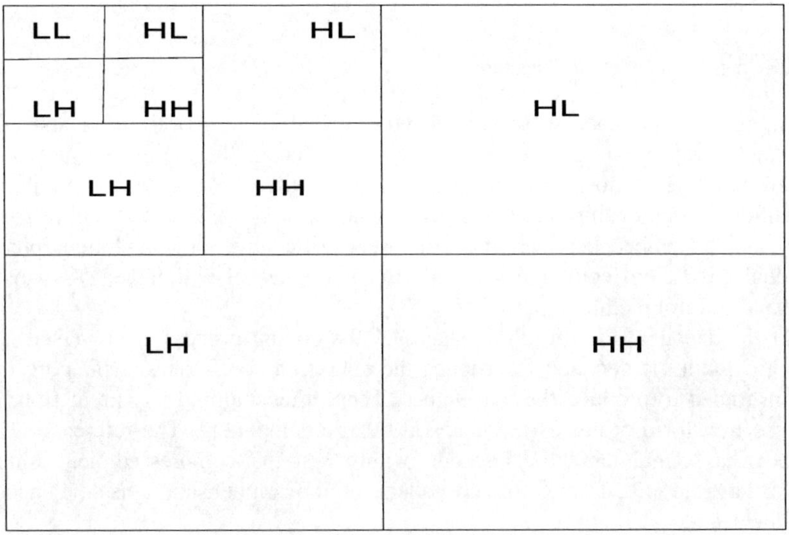

LL	HL	HL	
LH	HH		HL
LH		HH	
LH			HH

Fig. 7 Wavelet transform frequency decomposition [8]

Another view of wavelet transform is shown below:

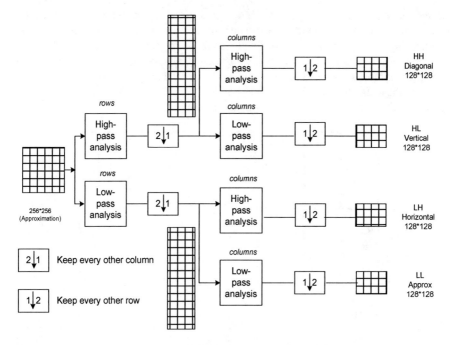

Fig. 8 First level of an image decomposition by wavelet sub-band filters

The number of samples in each resulting subband is as implied by the diagram above. The critical sub-sampling ensures that after each decomposition the resulting bands have one quarter of the samples of the input signal.

5.2.1 Wavelet Filters

The choice of wavelet filters has an impact on compression performance, filters having to have both compact impulse responses in order to reduce ringing artifacts and other properties in order to represent smooth areas compactly. It also has an impact on encoding and decoding speed in software. There are numerous filters supported by Dirac to allow a trade-off between complexity and performance. These are configurable in the reference software. These filters are all defined using the 'lifting scheme' for speed [12].

5.2.2 Padding and Invertibility

Clearly, applying an N-level wavelet transform requires N levels of subsampling, and so for reversibility, it is necessary that the dimensions of each component are divisible by 2^N. So if this condition is not met in the size of input image/frame, the input picture components are padded as they are read in, by edge values for best compression performance.

5.2.3 Parent-Child Relationships

Since each subband represents a filtered and subsampled version of the frame component, coefficients within each subband correspond to specific areas of the underlying picture and hence those that pertain to the same area can be related. It is most productive to relate coefficients that also have the same orientation in terms of combination of high- and low-pass filters. The relationship is illustrated below, showing the situation for HL bands i.e. those that have been high-pass filtered horizontally and low-pass filtered vertically.

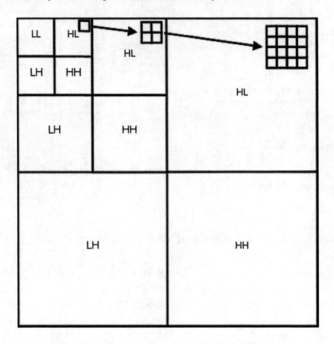

Fig. 9 Parent-child relationship between subband coefficients [8]

In the diagram it's easy to see that the subsampling structure means that a coefficient (the parent) in the lowest HL band corresponds spatially to a 2x2 block of coefficients (the children) in the next HL band, each coefficient of which itself has a 2x2 block of child coefficients in the next band, and so on. This relationship relates closely to spectral harmonics: when coding image features (edges, especially) significant coefficients are found distributed across subbands, in positions related by the parent-child structure, and corresponding to the original position of the feature.

These factors suggest that when entropy coding coefficients, it will be helpful to take their parents into account in predicting how likely, say, a zero value is. By coding from low-frequency subbands to high-frequency ones, and hence by coding parent before child subbands, parent-child dependencies can be exploited in these ways without additional signalling to the decoder [8].

5.3 Motion Estimation and Motion Compensation

5.3.1 Temporal Prediction Structures

Dirac encoder uses three types of picture.

a) Intra pictures (I pictures)
   These are coded without reference to other pictures in the sequence
b) Level 1 (L1) and Level 2 (L2) pictures
   Both of these are coded with reference to other previously coded pictures

The difference between L1 and L2 pictures is that L1 pictures are forward-predicted only (also known a P-pictures) whereas L2 pictures are B pictures are predicted from both earlier and later references.

The Dirac software employs a picture buffer to manage temporal prediction. Each picture is encoded with a header that specifies the picture number in display order, the picture numbers of any references and how long the picture must stay in the buffer. The decoder then decodes each picture as it arrives, searching the buffer for the appropriate reference pictures and placing the picture in the buffer. The decoder maintains a counter indicating which picture to 'display' (i.e. push out through the picture IO to the application calling the decoder functions, which may be a video player or may be something else). It searches the buffer for the picture with that picture number and displays it. Finally, it goes through the buffer eliminating pictures which have expired [8].

Nevertheless, the encoder operates with standard GOP modes whereby the number of L1 pictures between I pictures, and the separation between L1 pictures, can be specified; and various presets for streaming, SDTV and HDTV imply specific GOP structures. A prediction structure for picture coding using a standard GOP structure is shown below:

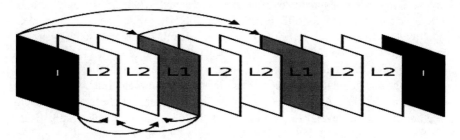

Fig. 10 Prediction of L1 and L2 pictures when L1 pictures are P pictures [8]

I-picture only coding

Setting the number of L1 pictures to be 0 on the encoder side implies that we don't have a GOP, and that we're doing I-picture only coding. I-picture only coding is useful for editing and other applications where fast random access to all pictures is required, but I-picture only coding is not essential for these applications with suitable support.

Interlace coding

Dirac supports interlace coding by coding sequences of fields, rather than frames.

5.3.2 Motion Estimation

Dirac uses hierarchical motion estimation that consists of three stages. In the first stage, motion vectors are evaluated for every block of each frame to one pixel accuracy using hierarchical motion estimation. In the second stage, these vectors are refined to sub-pixel accuracy. In the third stage, mode decision is performed in which motion vectors are aggregated by grouping blocks for similar motion. Motion estimation is most accurate when all three components are involved, but this is more expensive in terms of computation as well as more complicated algorithmically. Dirac uses the luma (Y) component only for ME.

5.3.2.1 Finding Motion Vectors of One Pixel Accuracy Using Hierarchical Motion Estimation

Hierarchical ME speeds things up by repeatedly down converting both the current and the reference frame by a factor of two in both dimensions, and doing motion estimation on smaller pictures. In hierarchical motion estimation, Dirac first determines the number of down conversion levels that can be calculated using equation 1 as follows:

$$level = \min\left(\log_2\left(\frac{width}{12}\right), \log_2\left(\frac{height}{12}\right)\right) \quad (1)$$

In equation 1, the number of down conversion levels is 4 and 3 for the CIF (352*288) and QCIF (176*144) frame format respectively. At each level, the process of down conversion reduces the height and width of the current and reference frame by a factor of 2 in each dimension. The size of the frame becomes one quarter at each level. Hence CIF image of size 352*288 and QCIF image of size 176*144 is reduced to 22*18 respectively in the last level.

The motion estimation is started from the lowest level resolution (level 4 in CIF) frame and gradually moved to higher level resolutions and finally reaches the original frame size. At each level of the hierarchy, except the smallest level, vectors from lower levels are used as a guide for searching at higher levels. Each block at the lower resolution level corresponds to four blocks at immediate higher resolution level, so each block at the lower resolution level provides a guide motion vector to at most 4 blocks at immediate higher resolution level. The block sizes are variable. The middle blocks are of size 12*12. The other block sizes are 10*10, 10*12, 10*6, 10*4, 10*8, 8*10, 6*10 or 4*10 depending upon the location of the block and the size of the frame and these are consistent at each level of the motion estimation hierarchy.

Survey of Dirac: A Wavelet Based Video Codec for Multiparty Video Conferencing

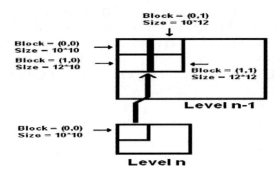

Fig. 11 Correspondence of one block at lower level to 4 blocks at immediate higher level

For motion estimation, candidate lists of motion vectors are generated at the lowest level. A candidate list consists of a number of vectors to be searched, which are centered at predicted MV and follows a square pattern at each level. The predicted MV can be zero MV, spatially predicted MV or guide MV. Spatially predicted MV may be zero MV, motion vector of previous horizontal block, motion vector of previous vertical block, the median of motion vectors of blocks 1, 2 and 3 for middle blocks and the mean of motion vectors of blocks 1 and 2 for last column blocks as shown in the Figure 10, depending upon the location of current block where motion estimation is performed. Guide MV is the best motion vector of a block at the lower level in the hierarchy and it is not available for lowest level.

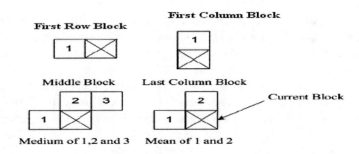

Fig. 12 Spatially Predicted Motion Vector of Dirac

Motion estimation is started at lowest level in the hierarchy. For lowest level, two candidate lists are generated, one is centered at zero MV and the other is centered at spatially predicted MV. For all other levels, three candidate lists are generated, two are same as above and the third list is centered at guide motion vector, which is the best motion vector of the corresponding block in the immediate lower level. Sum of Absolute Difference (SAD) is used for cost function. The vector which gives the minimum cost is recorded as best MV and the cost is recorded as best cost. The search pattern used at each level is square search with search range of ±5, ±4, ±3, ±2 and ±1 at levels 4,3,2,1 and 0 respectively in CIF. Thus there are at most 121, 81, 49 25 and 9 search points for each candidate list at levels 4,3,2,1 and 0 respectively.

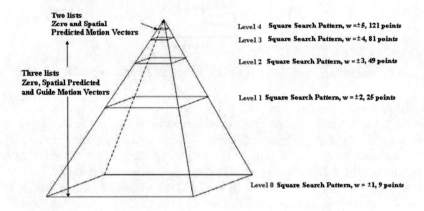

Fig. 13 Dirac Hierarchical Motion Estimation levels for CIF video format, where w is search range

5.3.2.2 Refine Pixel Accurate Motion Vector to Sub-pixel Accuracy

Pixel accurate motion vectors are refined to sub-pixel accuracy. Dirac provides motion vector accuracy upto ⅛ pixel, but ¼ pixel accuracy is default. Sub-pixel refinement process uses the pixel accurate motion vector as the initial guide for finding ½ -pel accuracy motion vectors. Similarly ½-pel accuracy motion vectors are used as a guide for ¼ -pel accuracy and ¼ -pel accuracy motion vectors are used as a guide for ⅛ -pel accuracy motion vectors, as illustrated in Figure 12. In order to achieve sub-pixel accuracy, the reference frame is up converted by 2, multiply the pixel accuracy motion vectors by 2 and search around the square window of ±1 to get motion vectors of ½ accuracy and so on.

Once pixel-accurate motion vectors have been determined, each block will have an associated motion vector (V0, W0). 1/2-pel accurate vectors are found by finding the best match out of (V0,W0) and its 8 neighbours: (V0+4,W0+4), (V0,W0+4), (V0-4,W0+4), (V0+4,W0), (V0-4,W0), (V0+4,W0-4), (V0,W0-4), (V0-4,W0-4). This in turn produces a new best vector (V1, W1), which provides a guide for 1/4-pel refinement, and so on until the desired accuracy. The process is illustrated in the figure below.

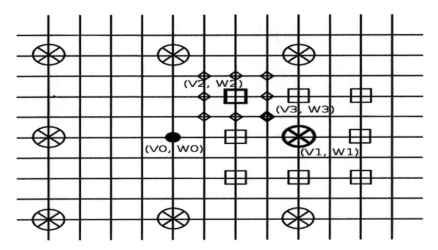

Fig. 14 Sub-pixel motion-vector refinement [8]

The sub-pixel matching process is complicated slightly since the reference is only up converted by a factor of 2 in each dimension, not 8, and so more accurate vectors require frame component values to be calculated on the fly by linear interpolation. This means that the 1/2-pel interpolation filter has a bit of pass-band boost to counteract the sag introduced by doing linear interpolation. It was designed to produce the lowest interpolation error across all the phases. If the best cost of the block, which is the cost until one pixel accuracy, is less than two times the multiplication of width and height of the block, then sets the same best motion vector by shifting left by precision (0 for 1 pixel accuracy, 1 for half pixel accuracy, 2 for quarter pixel accuracy and 3 for eight pixel accuracy). If the best cost is greater than two times the multiplication of width and height of the block, then the new cost is found by using prediction vector, which is the median of the neighboring vector. If the new cost is less than two times the multiplication of width and height of the block, then the best vector is set to prediction vector and best cost is set to the new cost. If the new cost is greater than two times the multiplication of width and height of the block, then sub-pixel refinement process of ½ and ¼ pixel accuracy motion vector starts.

5.3.2.3 Mode Decision by Using RDO Motion Estimation Metric

Mode decision is the last stage of motion estimation. At this point, block-level motion vectors for each frame, together with estimated costs for each are calculated. Dirac allows breaking up of a frame into sub-components for encoding. This is called 'Splitting'. Splitting mode is chosen by redoing motion estimation for the sub-Macro Blocks (MBs) and the MB as a whole, using RDO metric, suitably scaled to take into account the different sizes of blocks. Then the best prediction mode is chosen for each splitting mode. The RDO metric consists of a basic block matching metric, plus some constant times a measure of the local motion vector smoothness. The basic block matching metric used by Dirac is Sum of Absolute Differences (SAD). Given two blocks X, Y of samples, this is given by:

$$SAD(X,Y) = \sum_{i,j} |X_{i,j} - Y_{i,j}|$$

The smoothness measure is based on the difference between the candidate motion vector and the median of the neighboring previously computed motion vectors. The total metric is a combination of these two metrics. Given a vector V which maps the current frame block X to a block Y=V(X) in the reference frame, the metric is given by:

$$SAD(X,Y) + \lambda_{max}(|V_x - pred_x| + |V_y - pred_y|, 48)$$

The value λ is a coding parameter used to control the trade-off between the smoothness of the motion vector field and the accuracy of the match. When λ is very large, the local variance dominates the calculation and the motion vector which gives the smallest metric is simply that which is closest to its neighbors. When λ is very small, the metric is dominated by the SAD term, and so the best vector will simply be that which gives the best match for that block. For values in between, varying degrees of smoothness can be achieved. The parameter λ is calculated as a multiple of the RDO parameters for the L1 and L2 frames, so that if the inter frames are compressed more heavily then smoother motion vector fields will also result.

A macroblock (MB) consists of a 4x4 array of blocks and a sub-macroblock (sub-MB) consists of an array of 2*2 blocks. It means a Sub-MB consists of 4 blocks and a MB consists of 4 Sub-MB (16 blocks) as shown in figure.

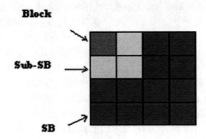

Fig. 15 Block, Sub-Macro Block, Macro Block

Total Frame size in blocks of CIF Format= 44 * 36

Total Frame size in Sub-MB of CIF Format= 22 * 18

Total Frame size in MB of CIF Format= 11 * 9

For prediction mode, we need costs for intra prediction and for bi-directional prediction. For a splitting mode we need costs for sub-macroblock and macroblock splittings for all four prediction modes. Dirac considers a total of 12 modes which consists of 3 Macro Block (MB) splitting levels and 4 prediction modes. A MB can be splitted in one of three ways.

Splitting level 0: no split, a single MV per reference frame for the MB;
Splitting level 1: split into four sub-macroblocks (sub-MBs), each a 2x2 array of blocks, one MV per reference frame per sub-MB;
Splitting level 2: split into the 16 constituent blocks, each have one MB.

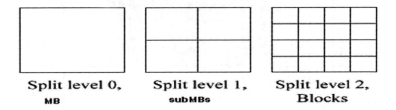

Fig. 16 MacroBlock Splitting Modes

There are four prediction modes available:

INTRA: intra coded, predicted by DC value;

REF1_ONLY: only predict from the first reference;

REF2_ONLY: only predict from the second reference (if one exists);

REF1AND2: bi-directional prediction, predicted from both first and second reference frame.

In motion estimation, an overall cost for each MB is computed, and compared for each legal combination of these parameters. This is a tricky operation, and has a very significant effect on performance. The decisions interact very heavily with those made in coding the wavelet coefficients of the resulting residuals, and the best results probably depend on picture material, bit rate, the block size and its relationship to the size of the video frames, and the degree of perceptual weighting used in selecting quantisers for wavelet coefficients.

5.3.3 Overlapped Block-Based Motion Compensation

Motion compensation in Dirac uses Overlapped Block-Based Motion Compensation (OBMC) to avoid block-edge artefacts which would be expensive to code using wavelets. Pretty much any size blocks can be used, with any degree of overlap selected: this is configurable at the encoder and transmitted to the decoder. One issue is that there should be an exact number of macro blocks horizontally and vertically, where a macro block is a 4x4 set of blocks. This is achieved by padding the data. Further padding may also be needed because after motion compensation the wavelet transform is applied, which has its own requirements for divisibility.

Dirac's OBMC scheme is based on a separable linear ramp mask. This acts as a weight function on the predicting block. Given a pixel p=p(x,y,t) in frame t, p may fall within only one block or in up to four if it lies at the corner of a block (see the figure below).

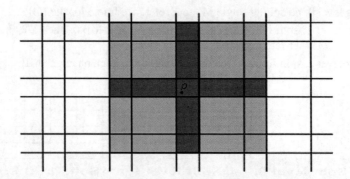

Fig. 17 Overlapping blocks. The darker-shades areas show overlapping areas.

Each block that the pixel p is part of has a predicting block within the reference frame selected by motion estimation. The predictor p for p is the weighted sum of all the corresponding pixels in the predicting blocks in frame t', given by $p(x-V_I, y-W_i, t')$ for motion vectors (V_i, W_i). The Raised-Cosine mask has the necessary property that the sum of the weights will always be 1:

$$\tilde{p}(x, y, t) = \sum w_i p(x - V_i, y - W_i, t), \sum w_i = 1$$

This may seem complicated but in implementation the only additional complexity over standard block-based motion compensation is to apply the weighting mask to a predicting block before subtracting it from the frame. The fact that the weights sum to 1 automatically takes care of splicing the predictors together across the overlaps.

As explained, Dirac provides motion vectors to sub-pixel accuracy, the software supports accuracy up to 1/8th pixel, although 1/4 pixel is the default. This means up converting the reference frame components by a factor of up to 8 in each dimension. The area corresponding to the matching block in the up converted reference then consists of 64 times more points. These can be thought of as 64 reference blocks on different sub-lattices of points separated by a step of 8 sub-pixels, each one corresponding to different sub-pixel offsets.

Sub-pixel motion compensation places a huge load on memory bandwidth if done naively, i.e. by up converting the reference by a factor 8 in each dimension. In Dirac, however, we just up convert the reference by a factor of 2 in each dimension and compute the other offsets by linear interpolation on the fly. In other words we throw the load from the bus to the CPU. The 2x2 up conversion filter has been designed to get the best prediction error across all the possible sub-pixel offsets.

5.4 Entropy Coding

Entropy coding is applied after wavelet transform to minimize the number of bits used. There are three stages in entropy coding used by Dirac in wavelet subband coefficient coding, that is as follows:

1) Binarisation
2) Context Modeling
3) Adaptive arithmetic coding

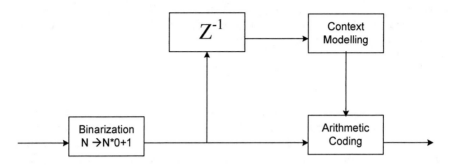

Fig. 18 Entropy Coding Block Diagram

The purpose of the first stage is to provide a bit stream with easily analyzable statistics that can be encoded using arithmetic coding, which can adapt to those statistics, thus reflecting any local statistical features.

5.4.1 Binarization

Binarization is the process of transforming the multi-valued coefficient symbols into bits. The resulting bit stream can then be arithmetically coded. Dirac uses an interleaved exp-Golomb binarisation.

In this scheme, add 1 to your value. Adding 1 ensures there will be be a leading 1. In binary form, the value will be a 1 followed by K other bits.

$$N + 1 = 1\ b_{k-1} \ldots \ldots b_0$$

These K bits ("info bits") are interleaved with K zeroes ("follow bits") each of which mean "another bit coming", followed by a terminating 1.

$$0\ b_{k-1} 0 \ldots \ldots 0\ b_0\ 1$$

Conventional exp-Golomb coding has the K zeroes at the beginning, followed by the 1 i.e 00...01bk-1bk-2 .. b0, but interleaving allows the decoder to run a single loop, two bits at a time, rather than in two loops.

5.4.2 Context Modeling

Context Modelling estimates the probabilites of symbols as they occur. It calculates the probabilities for 0 and 1 to be given to Arithmetic coding module. The context modelling in Dirac is based on the principle that whether a coefficient is small (or zero, in particular) or not is well-predicted by its neighbours and its parents. Therefore the codec conditions the probabilities used by the arithmetic coder for coding the first follow bit on whether the neighbouring coefficients or the parent coefficient are zero.The reason for this approach is that, whereas the wavelet transform largely removes correlation between a coefficient and its neighbours, they may not be statistically independent even if they are uncorrelated. The main reasons for this is that small and especially zero coefficients in wavelet subbands tend to clump together, located at points corresponding to smooth areas in the image, and are grouped together across subbands in the parent-child relationship. After binarization, a context is selected, and the probabilities for 0 and 1 that are maintained in the appropriate context will be fed to the arithmetic coding function along with the value itself to be coded [8].

5.4.3 Arithmetic Coding

Conceptually, an arithmetic coder can be thought of a progressive way of producing variable-length codes for entire sequences of symbols based on the probabilities of their constituent symbols. For example, if we know the probability of 0 and 1 in a binary sequence, we also know the probability of the sequence itself occurring. So if

P(0)=0.2, P(1)=0.8

then

P(11101111111011110101)=(0.2)3(0.8)17=1.8x10-4 (assuming independent occurrences).

Information theory then says that optimal entropy coding of this sequence requires $log_2 \left(\frac{1}{p}\right) = 12.4\ bits$. AC produces a code-word very close to this optimal length, and implementations can do so progressively, outputting bits when possible as more bits arrive. All arithmetic coding (AC) requires estimates of the probabilities of symbols as they occur, and this is where context modelling fits in. Since AC can, in effect, assign a fractional number of bits to a symbol, it is very efficient for coding symbols with probabilities very close to 1, without the additional complication of run-length coding. The aim of context modelling within Dirac is to use information about the symbol stream to be encoded to produce accurate probabilities as close to 1 as possible. Dirac computes these estimates for each context by maintaining a 16 bit probability word representing the probability of a zero value in that context. When a value has been coded, the probability is modified by incrementing (if zero was coded) or decrementing this value by a delta. The delta value itself depends only on the probability and is derived from a look-up table. If the probability is near 1 or 0, the delta is approximately 1/256; if the probability is near 1/2, the delta is approximately 1/32. This avoids the need to maintain explicit counts of 1 and 0 and makes for very efficient updating. There is no rescaling or division in the computation, and the estimate adapts rapidly for balanced probabilities and slowly for highly skewed probabilities, as it should.

The non-zero values in the higher frequency sub-bands of the wavelet transform are often in the same part of the picture as they are in lower frequency sub-bands. Dirac creates statistical models of these correlations and arithmetic coding allows us to exploit these correlations to achieve better compression. The

Survey of Dirac: A Wavelet Based Video Codec for Multiparty Video Conferencing 233

motion information estimated at the encoder also uses statistical modeling and arithmetic coding to compress it into the fewest number of bits. This compressed data is put into the bit stream, to be used by the decoder as part of the compressed video [8].

5.4.4 Dirac Bit Stream Syntax

Dirac bit-stream syntax structure is shown in Figure 17 (a ,b) for both intra and inter frames. It starts with sequence header. Sequence header starts with start of sequence code. The sequence header repeats itself after 10 frames for intra frame coding and after one group of pictures (GOP) for inter frame coding. Each sequence header and frame is treated as a parse unit.

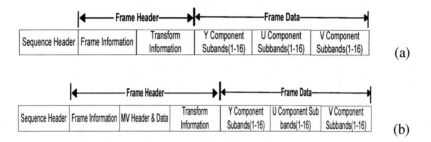

Fig. 19 Bit stream syntax (a) Intra Frame (b) Inter Frame

The first thirteen bytes of parse unit consists of start of parse unit (BBCD), parse code, next parse unit size and previous parse unit size as shown in Figure 18. The different parse codes related to start sequence, end of sequence, intra non reference picture arithmetic coding, intra reference picture arithmetic coding and inter reference picture arithmetic coding have been shown in Table 2.

Table 2 Parse code table

Parse Code	Description
0x00	Sequence header
0x10	End of sequence
0x0C	Intra Reference Picture (arithmetic coding)
0x08	Intra Non Reference Picture (arithmetic coding)
0x0D	Inter Reference Picture (arithmetic coding)

The 4 bytes of next parse unit represents the size of the current parse unit and 4 bytes of previous parse unit represents the size of previous parse unit. Both sizes should be exact otherwise the video will not be decoded.

Start of Parse Unit (BBCD)	Parse Code	No. of Bytes of Next Parse Unit	No. of Bytes of Previous Parse Unit
4 Bytes	1 Byte	4 Bytes	4 Bytes

Fig. 20 Parse unit first thirteen bytes

The input video has a certain frame rate. Each frame is encoded one by one in the same order as they are in the video. When a frame is being encoded the start of frame is the frame header. After frame header the frame data is divided into three components: Y, U and V. Each component has sub-bands. There are 16 sub-bands of each of Y, U and V components. The 16th band of Y component (Y16) is encoded first, then 15th and so on. After Y1, the last sub band of Y component, the U component will start. First U16------U1 and then V component will start. The YUV structure of a frame is as shown in Figure 19.

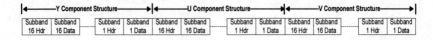

Fig. 21 YUV structure

Sequence header, frame header, bands header and bands data are all byte aligned. Except first thirteen bytes of each parse unit, Dirac generates bit-stream in interleaved exp-Golomb format. In this scheme, if a value N>=0 consists of k bits i.e. $x_{k-1}x_{k-2}...x_0$, then it is coded by adding 1 in N. The k bits are interleaved with K zeroes The bitstream will be $0x_{k-1}0x_{k-2}...0x_{01}$.

6 Comparison of BBC Dirac Video Codec

Dirac was developed through an evolution stage. Different versions of Dirac showed an incremental contribution of the previous versions. Currently the latest version of Dirac is 1.0.2.

6.1 Compression Ratio Test

The compression achieved by Dirac out performs H.264, however, this is only until a limit (approximately QF 10). Figure 20 and 21 shows that Dirac achieves better compression up to Quality Factor 10, thereafter, H.264 provides better compression.

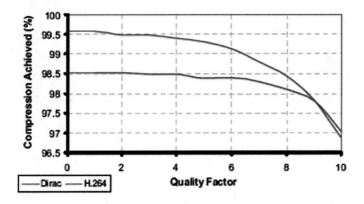

Fig. 22 Compression ratio comparison of Dirac and H.264 for "Container" QCIF sequence [1]

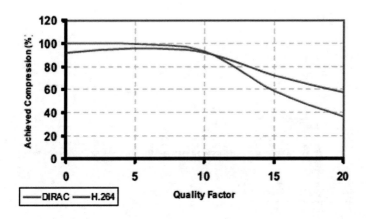

Fig. 23 Compression ratio comparison for "Riverbed" HD sequence [1]

6.2 PSNR Test

Figure 22 and 23 shows the PSNR comparison of Dirac and H.264 for "Container" QCIF and "Riverbed" HD sequences. For the QCIF sequence, H.264 maintains a higher PSNR at lower QF, however, there is marginal difference in PSNR between the codecs at QF 10. As for the HD sequence, the lack of CBR mode in the Dirac is evident after approximately QF 15 as the Dirac curve increases sharply while the H.264 curve starts to tail off.

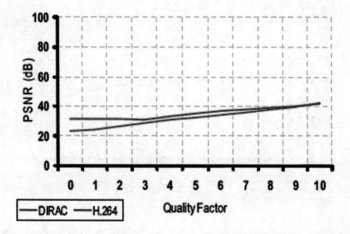

Fig. 24: PSNR comparison of Dirac and H.264 for "Container" QCIF sequence [1]

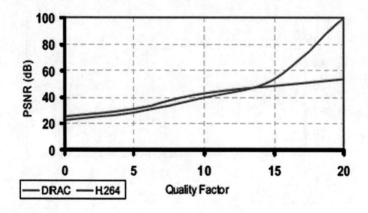

Fig. 25 PSNR comparison of Dirac and H.264 for "Riverbed" HD sequence [1]

6.3 SSIM Test

SSIM stands for Structural Similarity index. It is objective video quality metric. For a given reference video and a given compressed video, it is meant to compute a quality metric, based on perceived visual distortion. Figure 24 and 25 shows that there is substantial difference, and shows that Dirac is considerably worse at lower quality factor (bitrates). But Dirac reaches near to its maximum SSIM value at lower quality factors between 10-15.

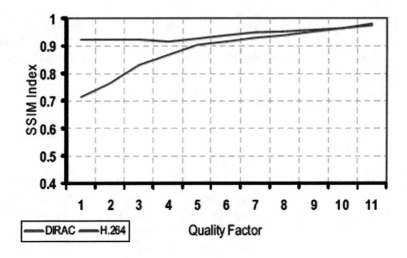

Fig. 26 SSIM comparison of Dirac and H.264 for "Container" QCIF sequence

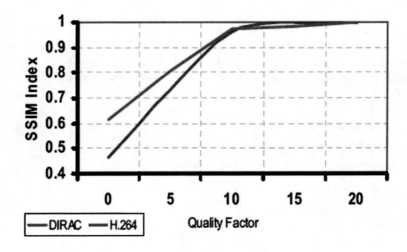

Fig. 27 SSIM comparison of Dirac andH.264 for "Riverbed" HD sequence [1]

7 Ongoing and Future Work Regarding BBC Dirac Video Codec

Efforts are going on across the world in order to improve the different modules of Dirac to make it suitable for real time video conferencing system. Different research papers have been published to address the computational complexity of the Dirac video codec. There is also working to embed new features in Dirac such as Scalability, Audio etc. Some of the ongoing work and future directions on Dirac video codec are now discussed.

7.1 Scalability

Scalability is an appealing concept. It can be seen as functionality where parts of a total bit-stream of a video are extracted as different sub-streams. Each sub-stream represents reduced source contents in temporal, spatial or SNR resolution as compared to the original bit-stream. The sub-streams are divided into layers i.e. base layer and enhancement layers. The base layer is a layer which is first encoded into sub-stream from the source content with low frame rate, low temporal or spatial resolution. The base layer is independently decodable and it produces an acceptable baseline quality. The residual information between base layer and source content is then encoded into one or more sub-streams and each sub-stream represents an enhancement layer. The enhancement layers cannot be independently decodable and can be added to the base layer to provide the user with high perceptual quality. The base layer should be reached at the end user correctly; otherwise the video will not be recovered properly. At decoder side, video can be decoded at any supported spatial, temporal or SNR resolution extracting a portion of the initial bit-stream. Scalability have been introduced in many video codecs that uses Wavelet Transform. In [16], Fully-scalable wavelet video coding using in-band motion compensated temporal filtering is described. Scalability allows great advantage in broadcasting/multicasting to users of heterogeneous capabilities.

There are three types of Scalability that is as follows:

  a) SNR Scalability
  b) Spatial Scalability
  c) Temporal Scalability.

7.1.1 SNR Scalability

SNR scalability involves generating two or more layers of the same spatiotemporal resolution but different video qualities from a single video source such that the base layer is coded to provide the basic video quality and the enhancement layer(s) when added back to the base layer reconstruct a higher quality reproduction of the input video. Since the enhancement layer is said to enhance the signal-to-noise ratio of the base layer, this type of scalability is called SNR scalability. This scalability is a tool intended for use in many video applications, for example, telecommunications and multiple quality video services with standard TV and High-Definition TV; Multi quality video on-demand services.

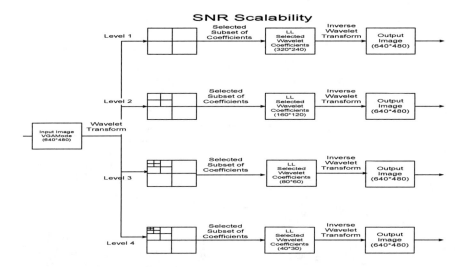

Fig. 28 SNR Scalability

7.1.2 Spatial Scalability

Spatial scalability involves generating two or more layers of the same spatiotemporal resolution but different video sizes from a single video source such that the base layer is coded to provide the basic video size and the enhancement layer(s) when added back to the base layer reconstruct a higher size reproduction of the input video. Since the enhancement layer is said to enhance the base layer for getting bigger size picture, this type of scalability is called Spatial scalability. In Wavelet Based Video Codecs, Spatial scalability allows you to reconstruct video of smaller size using fewer numbers of bands without compromising the video quality. It provides a tradeoff between the number of bands and video size.

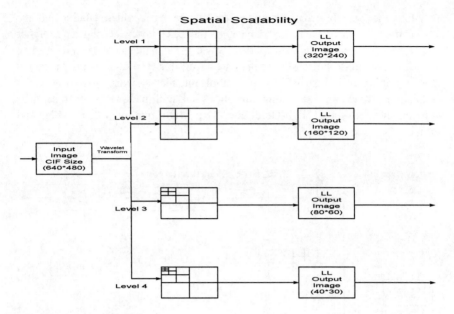

Fig. 29 Spatial Scalability

7.1.3 Temporal Scalability

Temporal Scalability allows you to reconstruct video using few numbers of frames. It is achieved by using bi-directionally predicted frames or B-frames. Temporal enhancement layers increase the frame-rate of the reproduced video by coding additional pictures.

7.1.4 Implementation Details in Dirac

In Dirac Wavelet In Dirac Wavelet Based Video Codec, the bit stream generated by encoder is splitted into 3 streams while keeping the header information in first stream as shown in *Figure 30*. Dirac encoder generates the encoded bit-stream of 16 bands, the Layer1 contains data for band 7-16 and all the header information of frame and sequence. It is independently decodable and is called base layer. Layer2 contains the bands from 4 to 6 and layer3 contains the bands from 1 to 3. Layer2 and layer3 are not independently decodable and are called enhancement layers. Each enhancement layer maintains the parse unit first thirteen bytes as shown in *Figure 20*. The layers of the bit stream for both intra nd inter frames after splitting are as shown in Figure 31(a,b). Note that the motion vector data and header information for inter frame are placed in the base layer.

Survey of Dirac: A Wavelet Based Video Codec for Multiparty Video Conferencing   241

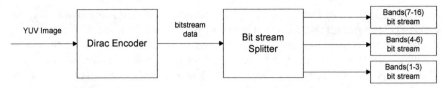

Fig. 30 Bitstream Splitting

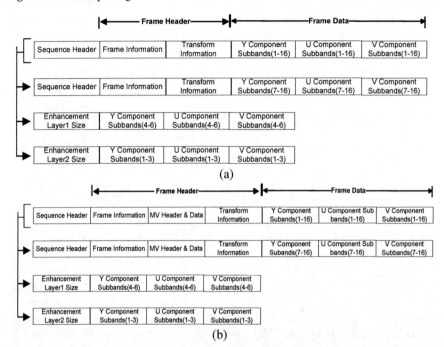

Fig. 31 Bit-stream Layers after Splitting (a) Intra Frame (b) Inter Frame

At the decoder side, the layers of bit-stream are combined and the header information for parse unit size of each parse unit is updated. The bit stream of different layers is joined as shown in *Figure 32*.

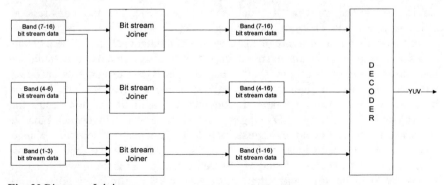

Fig. 32 Bitstream Joining

SNR scalability in Dirac allows to reconstruct video using fewer numbers of bands by compromising on video quality. It provides a tradeoff between the number of bands and video quality. For reconstruction video, the rest of the data is padding with zero. If all the bands are used, then there is no need of padding zero. The quality of reconstructed video with lesser number of bands is low as compared to greater number of bands. The bit stream structure after joining base layer and enhancement layer1 for both intra and inter frame for SNR scalability are as shown in Figure 33 (a,b). Note that the header information is necessarily appended; otherwise the bit stream will not be decoded.

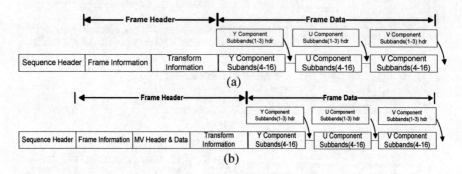

Fig. 33 Base layer and enhancement layer1 joining (a) Intra Frame (b) Inter Frame

For spatial scalability, the bit-stream generated by Dirac encoder is splitted into 3 streams same as SNR scalability. At the decoder side, the layers of bit-stream are combined and the header information for parse unit size of each parse unit is updated. The size of the reconstructed video and wavelet transform depth and related information is updated. Note that as oppose to the Figure 33(a,b), there is no need to append header information for the missing bands as the video is constructed from the specific number of 4-16 bands and its size is small.

7.2 Computational Complexity Reduction

To reduce the complexity of video encoder and improve the compression efficiency, fast and efficient motion estimation is crucial as it is the most expensive component in terms of memory and computational load. The existing ME search algorithm is too much expensive in terms of time. It uses multiple hierarchies for all types of inter (both P and B) frames. For example, to encode a CIF video format, the algorithm requires to generate 4 levels of hierarchies for both current and reference frames. It computes the optimum motion vector for each block by calculating SAD using the pattern as shown in Figure 13. Finally, the search is carried out at the original frame level. Obviously, it computes a huge amount of SAD calculations that take approximately 80% of the total encoding time. So, an efficient and faster ME search strategy is required, which could reduce the encoding time without affecting the accuracy. Efforts are going on to reduce the computational intensiveness by improving the ME algorithm. In [3], Tun et al. proposed a fast and efficient motion estimation algorithm which is based on semi hierarchical approach. It combines hierarchical and semi-hierarchical approaches for different types of inter frames. In [14], Khan et al. proposed a motion estimation algorithm for real-time H.264 video encoder, which is based on three dimensional recursive search (3DRS) algorithm. We have worked to improve the motion estimation module by replacing the current motion estimator with 3DRS [13]. In [4], we have implemented only the IPPP... structure. We have extended our approach to incorporate B frames. Now the structure will be IBBPBBP.... Like original 3DRS algorithm [13], the five candidate blocks involved in the initial prediction of motion vectors are as shown in Figure 34.

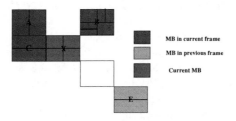

Fig. 34 Neighboring macro blocks used as candidates

The proposed Scheme is as shown in the Figure 35. The forward direction arrow in Figure 35 indicates that the temporal prediction MVs will be positive and a backward direction arrow indicates that they should be negative.

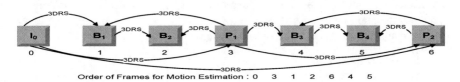

Fig. 35 Proposed Motion Estimation Scheme

It should also be noted that the best MVs from the P frames cannot be used directly as a temporal prediction since the prediction structure is different for both P and B as shown in Figure 35. The temporal prediction MVs for B frames requires scaling up or down since the temporal distances of both references are different for B frames. For example, the temporal distance of B1 to its 2nd reference is 2 and the temporal distance of P1 to its 1st reference is 3, so in order to use P1 MVs as a temporal candidate vectors, it should be multiplied by -2/3. −ve sign is used as the vectors are used for backward frame. Similarly, for B2 the MVs of P1 are multiplied by -1/3. It is important to note that the temporal predicted MVs are available only for the level 0 motion estimation and are uptil 1 pixel accuracy.

In order to reduce the level of complexity, the 3DRS algorithm is employed for all types of frames. The motion vectors are initialized with zero for those frames, whose reference frame motion vectors are not evaluated yet or whose reference frame is I frame for temporal prediction. For the remaining frames in the GOP, the 3DRS algorithm is invoked where initially the motion vector array of the reference frame is stored in memory. This array is then updated from left to right in a raster scan order. The best vector found is further refined by conducting the one pixel local diamond search around the position corresponding to the best motion vector. The best vectors found are up to one-pixel accuracy. The sub-pixel refinement and mode decision process is exactly the same as in the original Dirac algorithm.

7.3 Region of Interest Coding (Foveated Image Processing)

Region of interest coding allows you to code the certain portion of video with more resolution and the rest of the video portion with less resolution and hence help in efficient conservation of bit budget. This feature is already implemented in JPEG2000. Region of interest (ROI) coding is important in applications where certain parts of an image are of a higher importance than the rest of the image. In these cases the ROI is decoded with higher quality and/or spatial resolution than the background (BG). Example includes: Client/server applications where the server initially transmits a low quality/resolution version of an image. The client then selects an area of the image as a ROI and the server transmits only the data needed to refine of that ROI. Similarly, when browsing a digital photograph album it is often the case that we are looking for, or most interest in, the people/faces in those photographs. Using an automated face detection algorithm the region(s) of an image that contain faces can be coded as ROI's and therefore stored with more accuracy than non-face sub-images.

There are two approaches to Foveated Image Processing.

1) Object Based Foveation
2) Motion Based Foveation

7.3.1 Object Based Foveation

In this approach, we do Foveation, based on the object. Particular object is detected and then that region is sent with higher resolution and the remaining object is sent with lower resolution. So the main task is to detect the object. For example, in a video of a Football match, football is to be detected and that region where football is present would be sent with higher resolution and the other region would be sent with lower resolution.

7.3.2 Motion Based Foveation

In motion based Foveation, the region of interest is estimated based on its motion vectors. If motion vectors have a high value, means that there is more motion at that region so that region was sent with higher resolution and the remaining region was sent with lower resolution For example, in a football match the region in the image, where the football is present would generate high motion vectors and the remaining region would produce low value of motion vectors.

7.4 Video Conferencing System Based on Wavelet Transform

7.4.1 Association of Audio Stream

In communication, it is necessary an audio for developing a video conferencing system for meaningful delivery of video contents to the end users. The Dirac document does not mention any audio compression standard. The association of audio stream along with the video is vital. The Dirac video codec can be further improved by multiplexing the video and audio coded bit streams to create a single bit stream for transmission and de-multiplexing the streams at the receiving end [6].

7.4.2 Integrating Dirac in Openphone

In order to make Dirac an embedded codec, various features were modified. By following these steps and the demonstration that we have given in the form of Dirac, one could modify any standalone video codec to an embedded codec for any video conferencing system. The disk read/write operations are considered as being the most expensive ones, so we removed all the disk operations from Dirac and instead of reading the input video file and writing the output bit stream on a file, the embedded Dirac takes video input from the camera, encodes it using the traditional Dirac encoder and directs the output bit stream to a buffer. The output bit stream buffer is then handled by Open phone and it transmits the bit stream over network in order to send it to the receiver side. The timing diagram for the entire set of operations from grabbing at the local end to displaying at the remote end in the buffer I/O Open phone is illustrated in Figure 36.

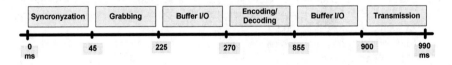

Fig. 36 Timing diagram for Dirac with buffer I/O [7]

At receiver side, the decoder receives the transmitted bit stream, decodes it using traditional Dirac decoder and again passes the decoded buffer back to Open phone, which displays the video on the local display window. In order to improve the frame rate of Dirac in Openphone, we modified some of the architectural behavior of Openphone. Original Openphone usually captures one frame, encodes and transmits it and move to the next frame. But Dirac usually generates a bitstream of 5 to 75 kilobytes which is impossible to fit in a couple of packets. So, we modified the Openphone architecture to transmit the whole bitstream of one frame before grabbing the next frame. After encoding, the transmission module is called and the bitstreams packets are sent until the whole bitstream for one encoded frame is transmitted. The new modified set of operations is presented in Figure 37.

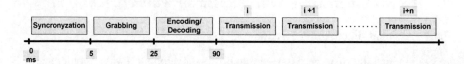

Fig. 37 Architectural modifications in Openphone [7]

By implementing these modifications we came up with a real time test bed in the form of Openphone. To check any video codec in real time video conferencing system, we can embed it Openphone and can calculate its bitrate, frame rate and other real time options [6].

References

[1] Onthriar, K., Loo, K.K., Xue, Z.: Performance Comparison of Emerging Dirac Video Codec with H.264/AVC. In: ICDT 2006 (September 2006)
[2] Ghanbari, M., Crawford, D., Fleury, M., Khan, E., Woods, J., Lu, H., Razavi, R.: Future Performance of Video Codecs. Video Networking Laboratory (November 2006)
[3] Tun, M., Loo, K.K., Cosmas, J.: Semi-Hierarchical Based Motion Estimation Algorithm for the Dirac Video Encoder. WSEAS Transactions on Signal Processing 4(5), 261–270 (2008)
[4] Ali, A., Ali, S.F., Khan, N.A., Masud, S.: Performance improvement in motion estimation of Dirac wavelet based video codec. In: Proc. ISCIT, Icheon, South Korea, pp. 764–769 (2009)
[5] Bradley, A.P., Stentiford, F.W.M.: JPEG 2000 and Region of Interest Coding. In: DICTA 2002: Digital Image Computing Techniques and Applications, Melbourne, Australia, January 21-22 (2002)
[6] Ravi, A.: Performance Analysis and Comparison of Dirac Video Codec With H.264 / Mpeg-4 Part 10 Avc
[7] Ahmad, A., Hussnain, M., Nazir, A., Khan, N., Masud, S.: Open Source Wavelet Based Video Conferencing System using SIP. In: Proceedings of International Conference on Information Society (I-Society 2010), London, UK, June 28-30 (2010)
[8] Dirac developer support, http://dirac.sourceforge.net/documentation/algorithm/algorithm/intro.htm
[9] Malvar, H.S., Hallapuro, A., Karczewicz, M., Korofsky, L.: Low-Complexity Transform and Quantization in H.264/AVC. IEEE Trans. on Circuits and Systems for Video Technology 13(7), 598–603 (2003)

[10] Eeckhaut, H., Schrauwen, B., Christiaens, M., Van Campenhout, J.: Speeding up the Dirac's Entropy Coder. In: Proceedings of 5th WSEAS Int. Conf. on Multimedia, Internet and Video Technologies, Corfu, Greece, August 17-19, pp. 120–125 (2005)
[11] Slides of Block diagram of DCT Based Codec, http://eeweb.poly.edu/%7Eyao/EL6123/videocoding.pdf
[12] Discrete Wavelet Transform, http://en.wikipedia.org/wiki/Discrete_wavelet_transform
[13] De Haan, G., Biezen, P.W.A.C., Huijgen, O.A.: True Motion Estimation with 3-D Recursive Search Block Matching. IEEE Transactions on Circuits and Systems for Video Technology 3, 368–379, 388 (1993)
[14] Khan, N.A., Masud, S., Ahmad, A.: A variable block size motion estimation algorithm for real-time H.264 video encoding. Elsevier Signal Processing: Image Communication 21, 306–315 (2006)
[15] Tun, M., Fernando, W.A.C.: An Error-Resilient Algorithm Based on Partitioning of the Wavelet Transform Coefficients for a DIRAC Video Codec. In: Information Visualization, July 05-07, vol. IV, pp. 615–620 (2006)
[16] Andreopoulos, Y., van der Schaar, M., Munteanu, A., Barbarien, J., Schelkens, P., Cornelis, J.: Fully-scalable wavelet video coding using in-band motion compensated temporal filtering. In: Proceedings of IEEE International Conference on Acoustics, Speech, and Signal Processing (ICASSP 2003), Hong Kong, China, vol. 3, pp. 417–420 (April 2003)

Author Index

Ali, Ahtsham 211
Ali, Syed Farooq 211

Bouthemy, Patrick 125

Davis, Michael 155

Gong, Shaogang 111

Hasegawa-Johnson, Mark A. 93
Hervieu, Alexandre 125
Huang, Thomas S. 93

Khan, Nadeem A. 211

Masud, Shahid 211
Mattivi, Riccardo 69
Mori, Taketoshi 195

Noguchi, Hiroshi 195

Odashima, Shigeyuki 195

Popov, Stefan 155

Sato, Tomomasa 195
Shao, Ling 69
Simosaka, Masamichi 195
Surlea, Cristina 155

Vatavu, Radu-Daniel 1

Yu, Hongchuan 21

Zhang, Jianguo 111
Zhang, Jian J. 21
Zhou, Huiyu 39
Zhou, Xi 93
Zhuang, Xiaodan 93